LABORATORY STATISTICS
Handbook of Formulas and Terms

LABORATORY STATISTICS

Handbook of Formulas and Terms

by

ANDERS KALLNER
Assoc. Professor (R)
Karolinska Univ. Laboratories
Stockholm SWEDEN

ELSEVIER AMSTERDAM · BOSTON · HEIDELBERG · LONDON · NEW YORK · OXFORD
PARIS · SAN DIEGO · SAN FRANCISCO · SINGAPORE · SYDNEY · TOKYO

Elsevier
225 Wyman Street, Waltham, MA 02451, USA
525 B Street, Suite 1800, San Diego, CA 92101-4495, USA

First edition

Library of Congress Cataloging-in-Publication Data
Kallner, Anders.
  Laboratory statistics : handbook of formulas and terms / by
  Anders Kallner. – First edition.
    pages cm
  Includes bibliographical references and index.
  ISBN 978-0-12-416971-5
1. Mathematical notation. 2. Mathematics–Terminology. I. Title.
  QA41.K32 2014
  519.503–dc23

                                                              2013019706

British Library Cataloguing in Publication Data
A catalogue record for this book is available from the British Library

For information on all **Elsevier** publications
visit our web site at store.elsevier.com

Printed and bound in USA
14 15 16 17 18 10 9 8 7 6 5 4 3 2 1

ISBN: 978-0-12-416971-5

Working together
to grow libraries in
developing countries

www.elsevier.com • www.bookaid.org

Contents

Acknowledgment

Extracts from the *International Vocabulary of Metrology— Basic and General Concepts and Associated Terms* (VIM), 3rd edition, JCGM 200:2012 are published with permission of the Director of the International Bureau of Weights and Measures (BIPM), in his functions as Chairman of the JCGM. The member organizations of the JCGM also retain full internationally protected right on their titles, slogans, and logos included in the JCGM's publications. The JCGM does not accept any liability for the relevance, accuracy, completeness, or quality of reproduced information and materials. The only official version is the original version of the document published by the JCGM.

Introduction

When we have to add, or multiply, even big numbers everything goes almost mechanically. This is a routine work, ..., the true mathematical thinking begins when one has to solve a real problem, that is to say, to identify a mathematical structure that would match the conditions of the problem, to understand principles of its functioning, to grasp connections with other mathematical structures, and to deduce the consequences implied by the logic of the problem. Such manipulations of structures are always immersed into various calculations since calculations form a natural language of mathematical structures. **Michel Heller (2008)**

This present "compendium" is for those who like me are engaged in practical laboratory work and do not have a major in statistical analysis and feel somewhat uncomfortable with the statistical jargon. We frequently face the need to analyze large amounts of data of various origins, collected for various purposes in routine or research work, and have discovered the power of spreadsheet programs in calculations and general data analysis.

Commercial statistical "packages" provide many of the analysis used in the laboratory. By necessity, the organization of the data in these packages has to accommodate many different requirements and is perhaps not optimal for a particular practical purpose. Laboratorians often desire to visualize their results graphically and interactively. The availability of spreadsheet programs has eliminated much of problems and hassle with calculations in statistics, provided simple understandable formulas are available. Indeed, simple spreadsheet programming can satisfy most of the necessary calculations and offer simple, efficient, and customized solutions.

This present compendium is not meant to be a "short course" in statistics but a source of a quick reference, repetition or explanation of formulas and concepts, and encourage development of statistical tools and routines in the research and routine laboratories.

Special attention has been given to expressions that can take different formats but, of course, give the same results. Exposing formulas in different formats may to some extent explain their origin, relation to other procedures, and their usage. We have tried to align formulas regarding style and terminology and group them in a logical order. Some formulas in the collection have been edited to facilitate applying in spreadsheet programs.

The selection of formulas in the compendium has developed during several courses in applied statistics for laboratorians and scientists with experimental projects. The number of worked examples is extensive and regularly enhanced by tables and figures. Whenever feasible the text makes reference to functions and routines in Microsoft EXCEL®.

Formulas have been collected and compared from many different sources, scientific literature, common textbooks, and the Internet. It is all out there, cast in different forms and shapes but may be difficult to find. An idea with this compendium is to have most of the statistical procedures used in the laboratory collected in one source and described in a standardized but not compressed format.

References to individual sources are not given but a list of contemporary literature.

A threat with preprogrammed routines is that, unless simple rules are violated and thus prevented from use, they will always produce an answer. The process of programming and calculating statistical routines has proved to deepen the understanding of the procedures and hopefully diminish erroneous use of established procedures. However, the author takes no responsibility for any erroneous decisions based on calculations using formulas in this compendium.

A comprehensive list of contents and an index facilitate the access of the desired concept or procedure.

VOCABULARY AND CONCEPTS IN METROLOGY

Many organizations have invested heavily in formulating internationally acceptable, clear, comprehensive, and understandable definitions of terms in metrology. Superficially, this may not seem to have any bearing on statistics. Basically, statistics is one way of formulating and expressing mathematical relationships, but we also need to agree on and use definitions of common concepts. The most extensive and internationally recognized list of concepts and their definitions is that created by the joint BIPM, ISO, IEC, IFCC, IUPAC, IUPAP, OIML, and ILAC document *International Vocabulary of Metrology—Basic and General Concepts and Associated Terms(VIM)*,downloadable at http://www.bipm.org/ (accessed 2013-06-30).

The definitions are reproduced *in extenso* from the VIM, but some notes have been deleted when pertaining to pure metrological problems.

The author is grateful for the interest and many excellent suggestions from students and other users of previous editions of the compendium. In particular, Professor Elvar Theodorsson, Department of Clinical Chemistry, University of Linköping, Sweden, has provided healthy critics.

Anders Kallner (anders.kallner@ki.se)

Some Notes on Nomenclature

Mathematical formulas may be difficult to decipher but are in fact unambiguous and comprehensive.

In this compendium, the formulas are not as compressed as they may be and therefore easier to understand. A few rules may help:

The number of items is abbreviated n or N.

$>$ is read "larger than," $<$ "smaller than," $\geq$ "larger than or equal to," $\leq$ "smaller than or equal to."

$\gg$ is read "much larger than," $\ll$ "much smaller than."

Fractions (division), a/b; multiplication $a \times b$.

Multiplications in the body of the text are written $\times$, i.e., $a \times b$.

Square root: $\sqrt{a}$ or explicit $\sqrt[2]{a}$, which allows for higher order roots.

Sum: $a_1 + a_2 + a_3 + \cdots + a_n$ is abbreviated: $\sum_{i=1}^{n} a_i$.

Sum of squares: $(a_1)^2 + (a_2)^2 + (a_3)^2 + \cdots + (a_n)^2$ is abbreviated $\sum_{i=1}^{n} a_i^2$.

A squared sum $(a_1 + a_2 + a_3 + \cdots + a_n)^2$ is $\left(\sum_{i=1}^{n} a_i\right)^2$.

Absolute value: $|a|$, i.e., disregarding any sign.

Standard deviation of a sample x is $s(x)$, $s(X)$, or s if there is no risk for misunderstanding.

Consequently, the standard error of the mean (SEM) is $s(\bar{x})$ or $s(\bar{X})$, as appropriate. The abbreviation SEM is also used.

The period (full stop) "." is used as the decimal sign and a comma "," as the 1000 separator.

Additional abbreviations as appropriate are explained in the text.

Some Greek letters used for certain purposes (small and capital):

Alpha: α and A, *Beta*: β and B, *Gamma*: γ and Γ, *Delta*: δ and Δ, *Epsilon*: ε and E, *Zeta*: ζ and Z, *Eta*: η and H, *Kappa*: κ and K, *Lambda*: λ and Λ, *My*: μ and M, *Xi;* ξ and Ξ, *Pi*: π and Π, *Rho*: ρ and P, *Sigma*: σ and Σ, *Tau*: τ and T, *Chi*: χ and X.

Formulas

BASICS

Logarithms and Exponents

The logarithm of a given number and a given base is the power to which the base must be raised to get the number.

If b is the base and a the given number, the logarithm is x. In many applications, the notation "log" refers to 10-logarithms (Briggs), i.e., the base 10 and ln refers to e-logarithms or "natural" logarithms with $e = 2.7183$ as the base:

$$\text{If } {}^b\log(a) = x, \quad \text{then anti} \log(x) = a = b^x \tag{1}$$

$$\text{thus if} {}^e\log(a) = x; \quad \ln(a) = x; \quad \text{then anti} \ln(x) = a = e^x \tag{2}$$

and

$$\text{if } {}^{10}\log(a) = x, \quad \text{then anti} \log(x) = a = 10^x \tag{3}$$

$$a \times b = c; \quad \log(a) + \log(b) = \log(c);$$
$$\frac{a}{b} = c; \quad \log(a) - \log(b) = \log(c) \tag{4}$$

$$\log(a^b) = b \times \log(a) \tag{5}$$

$$\frac{{}^c\log(a)}{{}^c\log(b)} = {}^b\log(a) \tag{6}$$

$$a^{-n} = \frac{1}{a^n} \tag{7}$$

$$a^{\frac{1}{n}} = \sqrt[n]{a} = \frac{1}{n} \times \log(a) \tag{8}$$

Microsoft EXCEL® commands: Natural logarithm: LN(a); antilog: EXP(LN(a)) (cf. 2).
10-logarithms (Briggs) LOG(a); antilog: $10^{\text{LOG}(a)}$ (cf. 3).
Value of $e=e^1$: EXP(1)=2.7183; e^b: EXP(b).

Examples

Let $a=5$, $b=10$, $c=3$, and $n=2$, then

$$^{10}\log(5) = 0.6990; \quad \text{anti}\log(0.6990) = 5 = 10^{0.6990}$$

Since $e=2.7183$ and $^e\log(5)=\ln(5)=1.61$; anti $\ln(1.61)=$ $5=e^{1.61}=2.7183^{1.61}$

$$5 \times 10 = 50; \quad \log(5) + \log(10) = \log(50);$$
$$0.6990 + 1 = 1.6990; \quad \text{anti}\log(1.6990) = 50$$

$$\frac{5}{10} = 0.5; \quad \log(5) - \log(10) = \log(0.5);$$

$$0.6990 - 1 = -0.3010; \quad \text{anti}\log(0.6990 - 1) = 0.5$$

$$\log(5^{10}) = 10 \times \log(5) = 6.990; \quad \frac{^3\log(5)}{^3\log(10)} = {}^{10}\log(5) = 0.6990$$

$$5^{-2} = \frac{1}{5^2} = \log(1) - 2 \times \log(5) = \text{anti}\log(0 - 2 \times 0.699)$$
$$= \text{anti}\log(-1.398) = \text{anti}\log(0.902 - 2) = 0.04$$

$$8^{1/3} = \sqrt[3]{8} = \frac{1}{3}\log(8) = \text{anti}\log\left(\frac{1}{3} \times 0.9031\right)$$

$$= \text{anti}\log(0.3010) = 2$$

Calculation of the logarithms, natural or 10-logaritms is directly available in spreadsheet programs. If mathematical tables or calculators are used, logarithms are conventionally expressed with four decimals to achieve sufficient precision for everyday use. Table values can be interpolated.

Derivation—Calculus

The derivative of a function at a given input value describes the best linear approximation of the function near that input value, i.e., the slope of the tangent in that point. Therefore, if the "first derivative" is set to zero and solved, the maximum(s) and/or minimum(s) of the function will be obtained. In higher dimensions, second, third, etc. derivatives can be calculated and if a second derivative is set to zero, the inflexion point of the original function is identified. The derivative of a function $f(x)$ is written dy/dx, y', or $f'(x)$ and interpreted as the "derivative of y with respect to x."

The partial derivative of a function of several variables is its derivative with respect to one of those variables while the others are held constant.

The partial derivative is written $\partial y/\partial x$.

Examples

The first derivative of a third degree function $y = \frac{1}{3}x^3 - 5x^2 - 11x - 5$ is $dy/dx = y' = f'(x) = x^2 - 10x - 11$ with maximum and minimum at $x = 5 \pm 6$, i.e., $x_1 = -1$ and $x_2 = +11$, respectively. The second derivative is $d^2y/dx^2 = y'' = f''(x) = 2x - 10$ and the inflexion point of the original function is $x = -5$.

Draw the three functions and confirm the maximum, minimum, and inflexion point!

If $y = n \times x^k + \text{constant}$, then a derivative will, in general terms, be

$$\frac{dy}{dx} = n \times k \times x^{(k-1)} \tag{9}$$

For a detailed discussion of derivative rules, derivatives, and partial derivatives, the reader is referred to special literature.

TRIGONOMETRY

Trigonometric Functions

In a right-angle triangle, i.e., a triangle with one angle equal to 90°, i.e., one side perpendicular to another side, the sides surrounding the right angle are called cathetus (*a* and *b* in

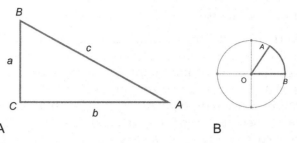

FIGURE 1   (A) Right-angle triangle. (B) The unit circle.

Figure 1A) and the opposite side the hypotenuse (c). The relation between these sides is expressed by the Pythagoras'
theorem:

$$a^2 + b^2 = c^2$$

The proportions or "image" of any triangle are determined
by the angles ($A=BAC$, $B=ABC$, and $C=ACB$). The angles can
be defined by the trigonometric functions referring to a right-
angle triangle (Figure 1A):

$$\sin A = \frac{a}{c}; \quad \sin B = \frac{b}{c}; \quad \cos A = \frac{b}{c}; \quad \cos B = \frac{a}{c}$$

$$\tan A = \frac{a}{b}; \quad \tan B = \frac{b}{a}; \quad \cot A = \frac{b}{a}; \quad \cot B = \frac{a}{b}$$

Provided the angle is known and expressed in *radians*
EXCEL provides numerical values of these quantities SIN(A),
COS(A), and TAN(A). The cotangent for an angle is the inverse
of its tangent and is not available as a separate function
in EXCEL.

Radian is defined as the angle *AOB* in the circle (Figure 1B)
where the arc *AB* is equal to the radius OB. Since the circumference is $2 \times$ radius $\times$ pi(π) corresponding to 360°, an angle of
1 radian will correspond to $360/(2 \times \pi)$ or 57.3°.

EXCEL provides conversions between degrees and radians:
RADIANS (angle in degrees) and *DEGREES* (angle in radians),
respectively. Therefore, to express the sine of 30°, the function
would be $SIN(RADIANS(30)) = SIN(0.52) = 0.5$. The reverse of
the trigonometric functions is *arcsine*, *arccosine*, and *arctangent*,

respectively. In EXCEL, the functions are *ASIN(A)*, *ACOS(A)* and *ATAN(A)*. Thus, to convert a sine of 0.5 to degrees, the function would be DEGREES(ASIN(0.5)).

Scales—Types of Data

Data can be expressed on four types or scales of data: nominal, ordinal, interval, and ratio.

Data on a nominal scale may be numbers or any other information that describes a property. There is no size relation between the entities.

Data expressed on an ordinal scale are of different sizes and can thus be ordered or ranked. The scale may be arbitrary and the intervals between numbers unequal. Data expressed on an ordinal scale can be measured and are thus quantities. Not all statistical procedures can be applied to ordinal data. Examples may be "good," "excellent," and "superior," or $+1, +2, +3$ etc. with no defined difference between the results.

Data with equal intervals between numbers are of two kinds and can be expressed on an interval scale and a ratio scale. The ratio scale is characterized by—apart from equally sized units—a natural zero value, whereas the interval scale may have an arbitrarily defined zero. A commonly cited quantity that is expressed on an interval scale is temperature expressed as degrees Celsius or Fahrenheit whereas if expressed in Kelvin a ratio scale is used. Consequently, 40 K is twice as much as 20 °C, whereas 40 °C is not twice as much as 20 °C. However, there are as many degrees between 40 and 20 °C as between 20 and 0 °C.

DISTRIBUTIONS OF DATA

Histogram

A histogram displays the number of data points in each of defined categories or intervals—often called "bins." It is a rough representation of the frequency probability distribution of data. The resolution and details of the distribution depend largely on the size and number of bins. Usually, the bin sizes are made equal in the interesting interval but that is not always the case. Designing a histogram manually is easy, but tedious and EXCEL offers two different possibilities. The simper is to activate the "Data analysis" function which is an add-in to the program and found under the Data tab. This is straightforward and allows an individual design of the bins as an option to those calculated by the program. The routine has the disadvantage of not allowing modifications interactively. A fully flexible procedure is obtained by the "frequency" function. This is an "array" function.

In short, define the desired bins, mark a set of empty cells, one cell more than the number of bins and write in the first cell =FREQUENCY(A1:AN₁, B1:BN₂) and press Control + Shift + Enter. The array is then created in the marked cells which are filled with the copied formula and subsequently with the number of items in each bin. The array formula will be the same in each of the marked cells. The bar graph can now be displayed. Any changes in the bins (B1:BN₂) or the data set (A1:AN₁) will immediately be reflected in the histogram.

The Normal and *t*-Distributions

The general concept is "probability density function." A special form, the Gauss distribution, is a symmetrical distribution around the most probable or frequent value, the mean, and a defined variability, the standard deviation. This distribution occurs if data are randomly distributed.

Gauss (normal) distribution:

$$G_{\mu,\sigma}X = \frac{1}{\sigma\sqrt{2\pi}} \times e^{-(x_i-\mu)^2/2\sigma^2} \tag{10}$$

The distribution is fully defined by two distribution-related constants, the population mean, μ, and the standard deviation, σ, and is graphically represented by the well-known bell-shaped curve, residing on a horizontal value axis (X-axis) and the frequency of observations on a vertical Y-axis. The peak of the curve is the average (16) of all observations belonging to the population and the variation or width of the distribution is related to the standard deviation (20). The standard deviation ($s(x)$) is a quantity value on the X-axis and formally obtained by solving the second derivative of the Gauss function when set equal to zero, i.e., an inflexion point of the function. Also see Figure 3 for an understanding of the standard deviation.

TABLE 1   The Relation between the Probability and the z-Value

Cumulative AUC (Probability)	z-Value ($s(x)$)	Remaining above the z-Value (1 – AUC)	Cumulative AUC (Probability)	z-Value ($s(x)$)	Remaining above the z-Value (1 – AUC)
0.500	0.00	0.500	0.99000	2.326	0.01000
0.841	1.00	0.159	0.99900	3.090	0.00100
0.975	1.96	0.025	0.99990	3.719	0.00010
0.977	2.00	0.023	0.99999	4.265	0.00001

Consequently, mean + 1 $s(x)$ represents a probability increase of (0.841 − 0.500) = 0.341, "the second $s(x)$" (0.977 − 0.841) = 0.136, and the mean ± 2 × $s(x)$ = 2 × (0.977 − 0.500) = 0.950. 3 $s(x)$ is usually approximated to a probability of 99.9% and 4 $s(x)$ to 99.99%.

The area under the curve (AUC) in the interval $-2\, x\, s(x)$ to $+2\, x\, s(x)$ is about 97.5% (see Table 1).

In the discussion of the properties of the normal distribution, a "standard normal distribution" is often used. This is characterized by a mean or average of 0 and a standard deviation of 1. Formula (10) is then simplified to

$$G_{0,1}X = \frac{1}{\sqrt{2\pi}} \times e^{-(x_i)^2/2} \tag{10A}$$

Although the shape is the same, the distribution's width and height may vary, also in relation to each other.

The cumulated AUC from $-\infty$ to z (a given number of standard deviations) is the cumulative probability (Table 1). This is coded as in EXCEL as *NORM.S.DIST(z,TRUE)*. *NORM.DIST (x,mean,standdev,TRUE)* will calculate the same but for any normal distribution.

In EXCEL, *NORM.S.DIST(z,FALSE)* and *NORM.DIST(x, mean,standdev,FALSE)* give the value of the normal distribution at the given value of z or x, respectively. These functions can be used to display a standard normal distribution curve (*NORM.S.DIST*) or any normal distribution (*NORM.DIST*) in EXCEL.

Calculation of the standard deviation (z-value) at which a certain probability (AUC) is reached is coded as *NORM.S.INV (probability)* for a standard normal distribution and *NORM. INV(probability,mean,standard deviation)* for the general case.

Skewness

This is a measure of the asymmetry of the probability distribution of a random variable. The skewness can be expressed numerically and is usually displayed in statistics software. A positive skewness indicates that the "tail" of the distribution is extended toward higher values (a "right skewness") and the majority of observations are found at lower values. Likewise, a negative skewness indicates a "left" skewness.

A normal distribution of a continuous variable has a skewness of 0.

There are many formulas to numerically estimate the skewness. Statistically, it was defined in terms of the second and third moment about the mean (Pearson) as

$$g = \frac{\frac{1}{n} \times \sum_{i=1}^{n} (x_i - \bar{x})^3}{\left[\frac{1}{n} \times \sum_{i=1}^{n} (x_i - \bar{x})^2\right]^{3/2}}$$

This formula can be expanded to compensate for the sample size:

$$G = \frac{\sqrt{n \times (n-1)}}{(n-2)} \times \frac{\frac{1}{n} \times \sum_{i=1}^{n} (x_i - \bar{x})^3}{\left[\frac{1}{n} \times \sum_{i=1}^{n} (x_i - \bar{x})^2\right]^{3/2}} =$$

$$\frac{n}{(n-1) \times (n-2)} \times \sum_{i=1}^{n} \left(\frac{(x_i - \bar{x})^2}{s}\right)^3 \tag{11}$$

which is the formula used in EXCEL *SKEW(cell A:cell B)*. There are tables of critical values of G for normal distributions.

By convention the skewness is interpreted as

- less than −1 or greater than +1: highly skewed.
- between −1 and −1/2 or between +1/2 and +1: moderately skewed.
- between −1/2 and +1/2: approximately symmetric.

There are different shortcuts to estimate the skewness. A crude estimate is the Bowley skewness or Quartile skewness, which gives an indication of the skewness. It compares the distances of the quartiles (25 and 75 percentiles, $p(0.25)$ and $p(0.75)$) from the median $p(0.50)$:

$$S = \frac{q_3 + q_1}{q_3 - q_1} = \frac{p(0.25) + p(0.75) - 2 \times p(0.50)}{p(0.75) - p(0.25)} \tag{12}$$

where q_3 is the distance between the $p(0.75)$ and median and q_1 the distance between $p(0.25)$ and the median. In this formula, $-1 \leq S \leq +1$. In a symmetrical distribution, when the $p(0.25)$ and $p(0.75)$ are equal, $S = 0$.

$S = \pm 0.1$ is regarded as a moderate skewness, whereas ± 0.3 is very noticeable.

Since only the middle two quartiles of the distribution are considered, and the outer two quartiles are ignored, this adds robustness to the measure but is also a caveat.

The choice of quartiles as the limits is arbitrary, and other pairs could be justified as well, e.g., $p(0.05)$ and $p(0.95)$.

The Pearson skewness is

$$S = \frac{\bar{x} - \text{mode}}{s} \tag{13}$$

and the Pearson second skewness

$$S_2 = \frac{\bar{x} - \text{median}}{s} \tag{14}$$

The "second skewness" may be more straightforward to estimate since the median can easily be calculated. The interpretation is to express the difference between the mean and median in terms of the standard deviation estimated as if the distribution were normal.

Example

The concentration of many components of blood is normally distributed (e.g., S-Calcium) but there are also many which distribution is skew, often positively, e.g., the concentrations of S-Triglyceride and S-Lp(a).

Suppose we have a data set that is normally distributed with a mean of zero [0] and a standard deviation of 1. Characteristically, the median (55), mean (16), and mode are equal. Suppose further that we add a number of results that are much larger than the mean above the mean and an equal number below, close to the mean. This causes the mean to move to a higher value and the standard deviation to increase, whereas the median remains the same. Consequently, the Pearson skewness (14) turns positive.

The "standard" deviation of a skew distribution can no longer be interpreted as 34 % of the distribution.

The "Pearson skewness index" is

$$S_k = \frac{3 \times (\text{mean} - \text{median})}{s} \tag{15}$$

An extensive example of skewness is given below.

Kurtois

This is a measure of the "peakedness" of the probability distribution of a real-value random variable. The kurtosis can be expressed numerically. Positive kurtosis (leptokurtic) indicates more peakedness than predicted by a normal distribution, and negative kurtosis (platykurtic) indicates less peakedness than a normal distribution. Zero [0] would indicate that the peakedness was as expected for a normal distribution. For instance, a rectangular distribution (83) has a kurtosis of [−1.2].

The t-Distribution

When the number of observations is low, the calculation of the normal distribution and parameters derived should be compensated for the low number. The compensation is in the calculation of the standard deviation by introducing the "degrees of freedom (df)" in the denominator of the calculation of the standard deviation and the population and sample standard deviation can be identified. See formulas (22) and (28). The t-distribution has the same bell-shaped symmetrical form as the normal distribution but is wider, the exact shape depending on the df (21). The *t-distribution* approaches that of the normal when the number of observations increases.

Transformation of Distributions

Many statistical evaluations require that the distribution of the data is, or is close to, normal. A data set can sometimes be transformed to approach a normal distribution, by recalculating the quantity values to logarithms, the square roots, or reciprocal values. These techniques reduce large values more than small values and positively skewed data sets may come close to normal. In general terms, reciprocal $(1/x_i)$ has a stronger effect on the skewness than the logarithmic transformation and the square root a weaker effect (see Table 2).

It is necessary to reestablish the original values before presenting the mean or variation of the original data set. The variation will usually no more be symmetrical around the mean value.

Transformation of skew distributions may have benefits as well as costs, e.g., loss of information.

TABLE 2   Descriptive Statistics of the Original and Transformed Distributions

	Orig	Ln	Sq Root
Mean (average)	2.57	0.87	1.57
Median	2.20	0.79	1.48
Standard deviation	1.08	0.38	0.32
Standard error of the mean (SEM)	0.12	0.04	0.03
Mode	1.70	0.53	1.30
Percentile 75	3.10	1.13	1.76
Percentile 25	1.77	0.57	1.33
	Orig	**Antiln**	**Squared**
Mean (average)	2.57	2.38	2.47
Median	2.20	2.20	2.20
Mean $-s$	1.49	1.62	1.58
Mean $+s$	3.66	3.50	3.57
CI ($\pm$SEM, $n=84$)	2.45-2.69	2.29-2.48	2.37-2.58
Mode	1.70	1.70	1.70
Percentile 75	3.10	3.10	3.10
Percentile 25	1.77	1.77	1.77

Example

The concentration of a series of 84 patient samples was measured.

Two data transformations were tried, logarithmic and square root (Figure 2). Both transformations resulted in a reduction of the skewness. In the lower panel of the table, the descriptive statistics of the transformed distributions have been transformed back to the original format. The means have shifted, and the confidence interval (CI) is no more symmetrical around the estimated mean; the nonparametric quantities (i.e., median, percentiles) are unchanged. As illustrated in Table 3, transforming using the reciprocal values is more powerful, i.e., reducing large values (above the median) proportionally more than small values, than the logarithmic transformation and the square root transformation less so.

Definitions and Metrics

In the following definitions,
x_i represents results of an observation and
n the number of observations.
Thus x_n is the nth observation of the series.

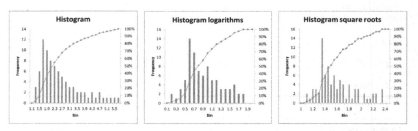

FIGURE 2   The appearance of a skew frequency distribution (left) and after transformations. The display of the distribution is related to the choice of bin sizes, whereas the skewness indicators are independent.

TABLE 3   Comparison of Skewness Estimates

	Orig	Ln	Sq Root	Reciproc
Skew (EXCEL)	1.16	0.49	0.83	0.22
Skew (Bowley)	0.36	0.23	0.29	0.09
Skew (Pearson-2)	0.34	0.21	0.28	0.03

Mean, arithmetic:

$$\bar{x} = \frac{\sum_{i=1}^{i=n} x_i}{n}; \quad \bar{x} = \frac{x_1 + x_2 + \cdots + x_n}{n} \tag{16}$$

The mean or average answers the question: *If all quantities had the same value, what would that be to achieve the same total sum?*

Example

The mean of observations $x_1=3$, $x_2=4$, $x_3=7$, and $x_4=10$
$\frac{\sum_{i=1}^{4} x_i}{n} = \frac{24}{4} = 6$. The sum of $6+6+6+6=24$.

The $\bar{x}$ is known as the sample mean or average, whereas the population mean or average is μ.

Mean, geometric:

$$\bar{x}_G = \pm\sqrt[n]{(x_1) \times (x_2) \times \cdots \times (x_n)}$$

$$\ln(\bar{x}_G) = \frac{1}{n}(\ln(x_1) + \ln(x_2) + \cdots + \ln(x_n)) \tag{17}$$

The geometric mean answers the question: *If all quantities had the same value, what would that be to achieve the same product?* In mathematical terms: "The *n*th root of the product of *n* numbers."

Example

The geometric mean of the above example is

$$\bar{x}_G = \pm\sqrt[4]{(3) \times (4) \times (7) \times (10)} = \pm\sqrt[4]{840} = \pm5.38$$

This expression may be more conveniently calculated using logarithms (1)–(8):

$$\ln(\bar{x}_G) = \frac{1}{4}(\ln(3) + \ln(4) + \ln(7) + \ln(10)) =$$

$$\frac{1}{4}(1.10 + 1.39 + 1.94 + 2.30) = 1.68$$

anti $\ln(1.68) = 5.38$

The product $5.38 \times 5.38 \times 5.38 \times 5.38 = 5.38^4 = 840$.

Note If the product of the results is negative, there must be an odd number of observations to give an interpretable result.

A consequence is that the antilogarithm of the arithmetic mean of a logarithmic distribution will be the geometric mean of the underlying distribution.

Mean, harmonic:

$$\bar{x}_h = \frac{n}{\sum_{i=1}^{i=n} \left(\frac{1}{x_i}\right)} \tag{18}$$

Example

Suppose you run the first half of a marathon in 5.00 km/h and the second in 9.00 km/h. The time would be the same (6.54 h) as if you had run the whole distance with 6.42 km/h; the harmonic mean 2/(1/5+1/9). If you ran 5.00 km/h for 3.27 h and 9.00 km/h for 3.27 h the average speed would be 7.00 km/h.

The harmonic mean will reduce the influence of extreme, large values but increase that of small values. The harmonic mean is equivalent to the inverse mean of the reciprocals of the values and a conveniently estimated mean of rates.

The harmonic mean is always the smallest of the three geometric, arithmetic, and harmonic means, the arithmetic always the largest and the geometric in between.

Mean, weighted:

$$\bar{x}_w = \frac{\sum_{i=1}^{i=k} (\bar{x}_i \times n_i)}{N} = \frac{\sum_{i=1}^{i=N} x_i}{N} \tag{19}$$

where $\bar{x}_i$ is the value of the quantity in each of the k bins, n_k is the number of observations in each of the corresponding bins, and N is the total number of observations, i.e., $N = \sum_{i=1}^{i=k} n_i$.

Example

A data set consists of five groups with 3, 3, 6, 2, and 7 items, $n=21$. The means of the groups were 5, 9, 3, 8, and 4, respectively:

$$\bar{x}_w = \frac{3 \times 5 + 3 \times 9 + 6 \times 3 + 2 \times 8 + 7 \times 4}{3 + 3 + 6 + 2 + 7} = \frac{104}{21} = 4.95$$

Standard deviation, sample:

$$s = \sqrt{\frac{\sum_{i=1}^{i=n}(x_i - \bar{x})^2}{n-1}} \qquad (20)$$

The standard deviation has the same dimension and unit as the measured quantity. In the graphical representation of the normal distribution—the familiar bell-shaped curve—it is the distance between the peak (arithmetic mean or average) and the first inflexion point where an increasing negative slope (above the mean) changes and begins decreasing towards horizontal. Since the function is symmetrical around the mean, the distance to the inflexion point on the other side of the mean is the same. The standard deviation is thus the positive second derivative of the normal frequency distribution (Gaussian distribution).

Note The standard deviation is the square root of a number and is always the positive root. If the standard deviation is used to describe, e.g., the interval where 95 % of observations of normally distributed data are expected, it is $\bar{x} \pm 1.96 \times s(x)$.

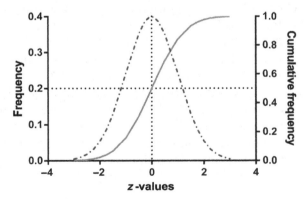

FIGURE 3 Frequency (hatched) and cumulative frequency (solid). z-Values are the number of standard deviations. The vertical dotted line represents average. The dotted horizontal line crosses the median (cumulative frequency 0.5) and intersects the gauss curve at its inflexion points, equal to the standard deviation.

TABLE 4 Frequency and Cumulative Frequency Mean$=0$, $S=1$

Probability	z-Value	Cumul Area	Frequency
0.001	−3.00	0.001	0.004
0.023	−2.00	0.023	0.054
0.159	−1.00	0.159	0.242
0.309	−0.50	0.309	0.352
0.500	0.00	0.500	0.399
0.691	0.50	0.691	0.352
0.841	1.00	0.841	0.242
0.975	1.96	0.975	0.058
0.977	2.00	0.977	0.054
0.999	3.00	0.999	0.004

In EXCEL, the function is *STDEV(cellA:cellB)*.

The "sum of squared" values appear in many statistical calculations (e.g., (20)). This means that the individual values are squared and the added together. Conveniently, there is a special function in EXCEL: *SUMSQ(cellA:cellB)*.

The standard deviation represents an interval on the X-axis and is expressed in the same unit as the quantity values. Due to the shape of the normal distribution curve, the "first standard deviation" counted from the mean will cover about 34.1 % of the AUC, the next only about 13.6 %, i.e., together about 48 %. The "third *s*" will cover only about 2.1 % of AUC. Therefore, the interval from −2 *s* to +2 *s* includes about 95.4 %, leaving about 2.5 % below −2 s and 2.5 % above +2 *s* (Tables 1 and 4). The z-values in Table 4 were calculated by EXCEL function *NORMSINV(probability)*. The cumulative area and frequency distribution are obtained by *NORM.S. DIST(z,TRUE)* and *NORM.S.DIST(z,FALSE)*, respectively.

df for sample standard deviation:

$$df = n - 1 \tag{21}$$

The subtraction of one [1] from the number of observations (*n*) in the calculations of the standard deviation can be explained by the data set being used once already to calculate the mean, thus losing one *df*. In general, the *df* are calculated as the sample size minus the number of estimated parameters. Therefore, the "corrected" variance is known as an unbiased estimator of the population variance; this correction is known as the Bessel correction.

To draw normal distribution curves in EXCEL

There are many ways to create a normal distribution curve in EXCEL. By defining the mean and standard deviation and a list of observations, the formula (10) can be used to calculate the frequency at each defined quantity value.

The frequency can also very conveniently be calculated using the function *NORMDIST(x,mean,s(x),FALSE)*. The function *NORM.S.DIST(z,FALSE)* will give the standard normal distribution, i.e., mean 0 and standard deviation 1. Exchanging "*FALSE*" for "*TRUE*" will give the cumulative frequency.

The functions *NORMINV* and *NORM.S.INV* will analogously calculate the frequency from the cumulative normal and standard normal distributions, respectively.

Standard deviation, short cut:

$$s = \sqrt{\frac{\sum_{i=1}^{i=n} x_i^2 - \frac{\left(\sum_{i=1}^{i=n} x_i\right)^2}{n}}{(n-1)}} = \sqrt{\frac{n \times \sum_{i=1}^{i=n} x_i^2 - \left(\sum_{i=1}^{i=n} x_i\right)^2}{n \times (n-1)}}$$

$$(22)$$

This formula will give identical results as Equation (20) but has the mathematical advantage of not calculating the individual differences between the observations and the mean which reduces possible rounding errors. It is advantageous to use in programming since calculation of the mean is not necessary. On the other hand, it involves the squares of x_i and the final difference must be calculated with a sufficient number of value digits to avoid undue rounding errors.

$$\text{Variance: } s^2 = \frac{\sum_{i=1}^{i=n} (x_i - \bar{x})^2}{n-1}$$

$$(23)$$

The variance occurs in many statistical calculations, e.g., propagation of errors and uncertainties and analysis of variance (ANOVA).

Example

Suppose we have a set of results 1,2,3,4,5,6,7,8,9. Calculate the standard deviation using formulas (20) and (22)!

The average is $45/9 = 5$. The degrees of freedom $9 - 1 = 8$.

Formula (20) yields, $\sum(x_i - \bar{x})^2 = 60$. The standard deviation is $\sqrt{60/8} = 2.74$.

The (22) generates: The sum of squared observations $\sum x^2 = 285$, the square of the sum of the observations $(\sum x)^2 = 2,025$, thus the $(\sum x)^2/(9) = 225$. The standard deviation is $\sqrt{(285 - 225)/8} = 2.74$

NB using EXCEL $SUMSQ(1,2...8,9) = 285$

The CI of the standard deviation of the population is

$$\sqrt{\frac{(n-1) \times s^2}{\chi^2_{(\alpha/2,\,(n-1))}}} = s \times \sqrt{\frac{(n-1)}{\chi^2_{(\alpha/2,\,(n-1))}}} \quad to \quad \sqrt{\frac{(n-1) \times s^2}{\chi^2_{(1-\alpha/2,\,(n-1))}}}$$

$$= s \times \sqrt{\frac{(n-1)}{\chi^2_{(1-\alpha/2,\,(n-1))}}}$$

(24)

Example

In an experiment, 20 results were obtained with an average of 5 g and a standard deviation of 0.7 g. Calculate the 95 % CI for the standard deviation!

First find the χ^2 (chi-square) values for the endpoints of the CI, i.e., 0.975 and 0.025 for the lower and higher limits, respectively. Remember that the χ^2 table and EXCEL command $CHIINV(\alpha,df)$ give the part of the distribution to the right (above) of the limit. Thus, for $df = (n-1)$, i.e., 19, $\chi^2_{(\alpha/2,\,(n-1))} = 32.9$; $\chi^2_{(1-\alpha/2,\,(n-1))} = 8.9$ and accordingly, the CI equals 0.53-1.02 g.

$$s \times \sqrt{\frac{(n-1)}{\chi^2_{(\alpha/2,\,(n-1))}}} = 0.7 \times \sqrt{\frac{19}{CHIINV(0.025,19)}} \quad to \quad s \times \sqrt{\frac{(n-1)}{\chi^2_{(1-\alpha/2,\,(n-1))}}} =$$

$$0.7 \times \sqrt{\frac{19}{CHIINV(0.975,19)}}$$

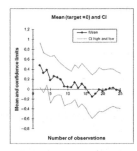 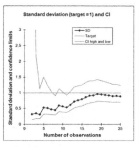 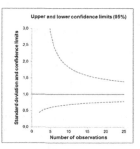

FIGURE 4   A set of 25 consecutive values were drawn from a normally distributed data set with a mean of 0 and standard deviation of 1. The mean and standard deviation were calculated from the first two, the first three, etc. values of the selection. As indicated, the estimated mean is comparatively stable and close to the target after about 10 observations, whereas the standard deviation levels out after about 20 observations. Note the CI is not symmetrical around the estimated standard deviation (middle panel). The rightmost panel shows the theoretical confidence interval of the standard deviation according to the formula (24).

Note The CI is not symmetrical around the standard deviation.

As illustrated in Figure 4, the CI of the standard deviation is very large if based on few observations. The same tendency is illustrated in a simulated example, whereas the CI of the mean includes the target value with a few observations.

The standard deviation according to Equation (20) underestimates s at low df, on an average, whereas the corresponding variance is correct. The reason is that taking the square root of the variances to compute the s numerically, reduces large variances more than small and therefore the mean of the s will underestimate the true population standard deviation (σ). This is particularly noticeable at small df. It can be corrected by the *c4-correction*. This is rarely applied mainly because the estimated variance is not liable to the same problem and the variance is used in Student's test and ANOVA, which are thus not affected.

The average underestimation of s is about 20 % with 2 observations, 8 % with 4 observations, and about 1 % with 25-26 observations. The *c4 correction* can be calculated using EXCEL:

$$c4 = EXP(LN(SQRT(2/(N-1))) + GAMMALN(N/2)$$
$$-GAMMALN((N-1)/2)) \qquad (25)$$

Number of Observations	2	3	4	5	10	25	26	100
$c4$	0.7979	0.8862	0.9213	0.9400	0.9727	0.9896	0.9901	0.9975

The correction is achieved by dividing the estimated $s(x)$ by the appropriate table value.

Standard deviation, population:

$$s_p = \sqrt{\frac{\sum_{i=1}^{i=n}(x_i - \bar{x})^2}{n}} \qquad (26)$$

The sample standard deviation (20) approaches the s_p if the number of observations is large and equals the root mean square (27) if $\bar{x}$ is zero [0].

If the mean is set to zero, by definition (e.g., evaluating differences of quantity values in a quality control experiment or in a regular wave function, e.g., the sinus function), the *df* is equal to the number of observations since no mean has been calculated.

s_p is also known as the biased population standard deviation.

Root mean square (mean, quadratic; RMS):

$$x_q = \sqrt{\frac{\sum_{i=1}^{i=n}x_i^2}{n}} \qquad (27)$$

The RMS is a measure of the magnitude of a varying quantity. It is especially used when variables are positive and negative, e.g., in describing a sinus wave. It corresponds to the standard deviation of a number of observations with a mean of zero [0] (cf. 26).

The relation between the quadratic mean and the average of a set of normally distributed values with a population standard deviation of $s_p(x)$ (26) is

$$x_q^2 = (\bar{x}_i)^2 + s_p^2 \qquad (28)$$

The RMS of the error of a measurement (Δ) (RiLiBAEK IQC rules in Germany).

The root mean square is used in quality management as a metric that addresses the imprecision *and* bias. The root mean square of the deviation of a measurement is

$$\Delta = \sqrt{\frac{\sum_{i=1}^{i=n}(x_i - x_0)^2}{n}} \tag{29}$$

where x_i is the observations and x_0 a "true value."

If the mean (16) of the observations and their standard deviation (20) are ($\bar{x}$) and (s), respectively and the systematic deviation ($\bar{x} - x_0$) = δ then, since $\sum_{i=1}^{n}(x_i - x_0)^2 = \sum_{i=1}^{n}(x_i - \bar{x})^2 + n(\bar{x} - x_0)^2$ (29) can be rewritten as

$$\Delta = \sqrt{\frac{\sum_{i=1}^{i=n}(x_i - x_0)^2}{n}} = \sqrt{\frac{\sum_{i=1}^{i=n}(x_i - \bar{x})^2 + n(\bar{x} - x_0)^2}{n}} =$$

$$\sqrt{\frac{(n-1)s^2}{n} + \delta^2} \tag{30}$$

Standard deviation, relative (coefficient of variation):

$$CV = \frac{s}{\bar{x}} \tag{31}$$

Standard deviation, relative, percent:

$$\%CV = \frac{100 \times s}{\bar{x}} \tag{32}$$

The coefficient of variation, percent, is often abbreviated %CV.

Example

A series of results were 3, 3, 6, 2, 7, 5, 9, 3, 8, and 4. The mean is 5 and the standard deviation 2.40. The %CV is then $100 \times 2.401/5 = 48.1$ %.

Note The relative standard deviation (RSD) and the coefficient of variation are always (by definitions 31 and 32) calculated from one [1] standard deviation.

Standard deviation, pooled:

$$s_{pool} = \sqrt{\frac{(n_1 - 1) \times s_1^2 + (n_2 - 1) \times s_2^2 + \cdots + (n_k - 1) \times s_k^2}{(n_1 + n_2 + \cdots + n_k) - k}} =$$

$$\sqrt{\frac{\sum_{i=1}^{k} \left((n_i - 1) \times s_i^2\right)}{\sum_{i=1}^{k} n_i - k}} \qquad (33)$$

The pooled standard deviation is used to find an estimate of the population standard deviation given several different samples and their number and standard deviation measured under different conditions. The mean of the different series of measurements may vary between samples, but the standard deviation (imprecision) should remain almost the same. Pooling standard deviations may provide an improved estimate of the imprecision.

The pooled standard deviation can be used to describe the total performance of a laboratory which uses many instruments to measure the concentration of the same analyte.

Example

A laboratory used three instruments to measure the concentration of the same analyte. The first instrument measured 12 samples with a standard deviation (s) of 0.35, the second 3 samples with $s = 0.55$, and the third instrument measured 7 samples with $s = 0.40$. Estimate the pooled standard deviation, representing that for samples analyzed on a random choice of instruments.

$$s_{pool} = \sqrt{\frac{(12 - 1) \times 0.35^2 + (3 - 1) \times 0.55^2 + (7 - 1) \times 0.40^2}{(12 + 3 + 7) - 3}} =$$

$$\sqrt{\frac{2.91}{22 - 3}} = \sqrt{0.15} = 0.39$$

Squared, multiplied by $(N - k)$, and applied to repeated measurements of the same material, this is identical to the "within group" mean square of the ANOVA; compare the formula for within-series sum of squares in ANOVA analysis, SS_w (105).

If the number of observations in each group is the same $(n_1 = n_2 = \cdots = n_i)$, then

$$s_{\text{pool}} = \sqrt{\frac{s_1^2 + s_2^2 + \cdots + s_k^2}{k}} = \sqrt{\frac{\sum_{i=1}^{k} s_i^2}{k}} \tag{34}$$

i.e., the s_{pool} is the square root of the average *variance* of the groups.

If the imprecision is not constant but proportional to the measured quantity, it is reasonable to assume that there is a constant RSD, characterizing the imprecision. The calculated RSD can be calculated from the results of a series of measurements as

$$\text{RSD}_{\text{pool}} = \sqrt{\frac{\left(\sum_{i=1}^{k} \frac{(n_i - 1) \times s_i^2}{\bar{x}_i^2} \right)}{\sum_{i=1}^{k} n_i - k}} \tag{35}$$

If there are only two observations in each group, then Equation (33) can be rearranged to

Standard deviation (Dahlberg):

$$s = \sqrt{\frac{\sum_{i=1}^{N} (x_{1i} - x_{2i})^2}{2 \times N}} = \sqrt{\frac{\sum_{i=1}^{i=N} d_i^2}{2 \times N}} \tag{36}$$

where d is the difference between results of duplicate measurements, N is the number of pairs (duplicates). The standard deviation estimated by the "Dahlberg" formula assumes that the standard deviation is constant in the measuring interval (homoscedastic). In any case, it will produce a value that considers all data pairs.

To estimate the RSD from duplicate observations, using the "Dahlberg formula," the relative difference between each pair of observations is calculated:

$$\text{RSD} = \sqrt{\frac{\sum_{i=1}^{i=N} \left(\frac{2 \times (x_{i1} - x_{i2})}{(x_{i1} + x_{i2})} \right)^2}{2 \times N}} \tag{37}$$

Example

A series of samples were measured in duplicates: $3.6 - 4.0$; $10.3 - 9.8$; $5.2 - 5.9$; $4.6 - 4.2$; and $7.2 - 7.6$. Calculate the differences between the duplicates: -0.4; 0.5; -0.7; 0.4; and -0.4. The sum of the squared differences $d_1^2 = 1.22$; $s_1 = \sqrt{\dfrac{1.22}{2 \times 5}} = 0.35$.

Similarly, the RSD:

$$RSD = \sqrt{\frac{\left(\dfrac{-0.4}{3.8}\right)^2 + \left(\dfrac{0.5}{10.75}\right)^2 + \left(\dfrac{-0.7}{5.55}\right)^2 + \left(\dfrac{0.4}{4.4}\right)^2 + \left(\dfrac{-0.4}{7.7}\right)^2}{2 \times 5}}$$

$$= \sqrt{\frac{0.040}{2 \times 5}} = 0.063 \text{ or } 6.3\,\%$$

The %CV calculated from the s_1, with the mean of 6.24, the % RSD $= 5.6\,\%$.

Standard deviation, geometric is obtained by calculation of the standard deviation from logarithmically transformed measurement results $\bar{x}_g$ and s_g, respectively. The logarithmic CI is thus as follows:

$$CI_g = \bar{x}_g \pm z \times \frac{s_g}{\sqrt{n}} \tag{38}$$

To get the quantity values the antilogarithms of $\bar{x}_g$ and s_g must be calculated.

Standard error of the mean, SEM:

$$SEM = s(\bar{x}) = \frac{s}{\sqrt{n}} = \sqrt{\frac{\sum_{i=1}^{i}(x_i - \bar{x})^2}{n \times (n - 1)}} \tag{39}$$

The standard error of the mean expresses the interval within which a repeated estimate of the mean is assumed to occur with a given probability. The metric thus expresses how well a mean is estimated. This is different from the standard deviation (s) which describes the width of the distribution and thus participates in the definition of the normal distribution. The standard error of the mean is also abbreviated SEM $= s(\bar{x})$, and with the same logics, the standard deviation is specified and abbreviated $s = s(x)$.

Confidence interval, CI:

$$CI = \bar{X} \pm t_{(1-\alpha;n-1)} \times \frac{s}{\sqrt{n}} \tag{40}$$

The magnitude of the t-value depends on the degrees of freedom ($df=n-1$) and the level of confidence ($1-\alpha$). The t-value should be taken from a Student's t-value table or calculated in EXCEL by the function $TINV(1-\alpha,df)$. For large numbers of observations (often suggested >30), the t-value can be obtained from a "z-table" or by EXCEL: $NORMSINV(1-\alpha)$, i.e., the t-distribution approaches the normal distribution.

Thus, for large numbers $t_{(1-\alpha,n-1)} \approx z_{(1-\alpha)}$.

Proportions

The binominal distribution describes the distribution of values that can only have two outcomes, e.g., healthy-nonhealthy or heads and tails in flipping a coin.

If the total number of observations (trials) is n and the number of successes (hits, positive findings, etc.) is r, then the probability ($0 \leq p \leq 1$) or proportion of successes in each independent trial is

$$p = \frac{r}{n} \tag{41}$$

The variance of the *binominal distribution* is $[p \times (1-p)]$ and thus the

Standard deviation of a proportion ($p/(1-p)$):

$$s(p) = \pm\sqrt{p \times (1-p)} \tag{42}$$

The frequency distribution of a proportion is the binominal distribution, but usually, the normal approximation to the normal distribution can be used. This requires a sufficient number of observations expressed and sufficient in this case requires that both $n \times p$ and $n \times (1-p)$ exceed 5 which is equal to requiring that r and $n-r$ are above 5.

Standard error of a ratio (proportion [p])

$$s(\bar{p}) = \sqrt{\frac{p \times (1-p)}{n}} \tag{43}$$

where p is the proportion and n is the number of observations, i.e., the sample size.

The CI

$$CI(\bar{p}) = z \times \sqrt{\frac{p \times (1-p)}{n}} \qquad (44)$$

The above method is often referred to as the traditional method to estimate $s(\bar{p})$. It is not applicable if the proportion is large or small; often limits of 0.1 and 0.9 are specified. It is thus often inappropriate to use the traditional method for estimating the standard error and the CI of diagnostic sensitivity, diagnostic specificity, and prevalence of disease, where proportion below and above the limits, respectively, is common. Instead, the following procedure is recommended, often referred to as Wilson's method.

If r is the number of observations that refer to a particular property (e.g., testing positive in an investigation among diseased), TP in a sample of n observations (e.g., diseased), then

$$p = \frac{r}{n} \quad \text{and} \quad q = 1 - p \qquad (45)$$

Calculate:

$$A = 2 \times r + z^2 \qquad (46)$$

$$B = z \times \sqrt{z^2 + 4 \times r \times q} \qquad (47)$$

$$C = 2 \times (n + z^2) \qquad (48)$$

where z is 1.96 to correspond to a 95 % CI.

Then, the CI is as follows:

$$\frac{A - B}{C} \quad \text{to} \quad \frac{A + B}{C} \qquad (49)$$

Example

In a group of 85 (n) known diseased, only 5 tested positive (r) with a new test. Estimate the proportion (p) of true positives ($TP = sensitivity$) and its CI using traditional and alternative methods!

Thus, $n=85$ and $r=5$ and $p = \frac{5}{85} = 0.059$; $n \times (1-p)=80$; $n-r=80$:

$$A = 2 \times 5 + 1.96^2 = 13.8$$

$$B = 1.96 \times \sqrt{1.96^2 + 4 \times 5 \times \left(1 - \frac{5}{85}\right)} = 9.3$$

$$C = 2 \times \left(85 + 1.96^2\right) = 177.7$$

and

$$\frac{13.8 - 9.3}{177.7} = 0.025 \text{ to } \frac{13.8 + 9.3}{177.7} = 0.130$$

Thus, the proportion is 6 % with a CI of 2.5-13 %.

Note The CI is not symmetrical around the proportion.

The CI estimated according to the "traditional method" (44) is $\pm 1.96 \times \sqrt{\frac{\frac{5}{85} \times \left(1 - \frac{5}{85}\right)}{85}} = 0.052$, thus from 0.007 to 0.11 or 0.7 % to 11 %.

At low proportions, the CI with this approach may produce impossible negative numbers of the probability. Try a positive diagnostic sensitivity of 3 %! (The confidence limit is negative for $p < 3.7$ and larger than 1, for $p > 96.3$).

Uncertainty of the difference between two proportions
The general rule of error propagation is applicable:

$$u(p_1 - p_2) = \sqrt{\left(s(\bar{p}_1)\right)^2 + \left(s(\bar{p}_2)\right)^2} =$$
$$\sqrt{\left(\frac{p_1 \times (1 - p_1)}{n_1}\right)^2 + \left(\frac{p_2 \times (1 - p_2)}{n_2}\right)^2} \quad (50)$$

z-score:

$$z = \frac{x_i - \bar{x}}{s} \quad (51)$$

The *z-score* thus positions an observation in relation to the mean and the distribution; the position is expressed in standard deviations. The 95 % confidence limit would thus be expressed as $\pm 1.96\ z$. The z-score is often used to express the performance of measurements by different procedures and different quantity values. The probability that a value should be outside (larger than) the z-score is $1 - NORMSDIST(z)$. For a two-sided evaluation, use $2 \times (1 - NORMSDIST(z))$ or a normal t-table.

Poisson Distribution

This describes independent and random occurrences of events, e.g., radioactive decays per time unit [frequency] (*dpm*).

If n is the number of events per time unit (e.g., *dpm*) and T is the number of time units the total number of occurrences is

Number of events:

$$\mu = n \times T \qquad (52)$$

Standard deviation of the Poisson distribution:

$$\pm\sqrt{\mu} \qquad (53)$$

Coefficient of variation of the Poisson distribution:

$$\pm\frac{\sqrt{\mu}}{\mu} = \pm\frac{1}{\sqrt{\mu}} \qquad (54)$$

Example

The activity of radon was found to be 340 Becquerel/m^3 (1 Becquerel = 1 disintegration/s). The standard deviation is ±18 Becquerel/m^3 and the RSD 0.05 or %CV = 5 %.

ROBUST ESTIMATORS

Median (50-Percentile)

The number that separates the higher half of a sample, a population, or a probability distribution from the lower half represents the median. The median of a finite list of numbers can be found by arranging all the observations from the lowest value to the highest value and picking the middle one. If there is an even number of observations, the median is not unique, so one often takes the average of the two middle values.

In a series of n odd-ordered numbers:

$$\text{median} = x_{(n+1)\times0.5} \qquad (55)$$

In a series of n even-ordered numbers:

$$\text{median} = \frac{x_{(n \times 0.5)} + x_{(n \times 0.5 + 1)}}{2} \tag{56}$$

Mode

The mode of a discrete probability distribution is the value x at which its probability function takes its maximum value or peak. In other words, it is the value that is most likely to be sampled. A density function may have several peaks and is then referred to a multimodal in contrast to unimodal.

In a normal distribution the mode, median and mean coincide.

Percentiles

There is no universally accepted definition of a percentile and statistical software packages may use different techniques for its calculation. The problem is that of definition and rounding: Thus, the percentile can either be defined as the lowest score that is above the percentage looked for or the smallest score that is greater than—or equal to—the percentage. This problem is particularly important if the number of observations is small.

EXCEL: PERCENTILE(interval,p) where p is the percentile, e.g., 0.75 for the 75 percentile.

Example

The following procedure calculates the score of the value nearest below the percentile and interpolates in the interval between the numbers closest to the percentile. Consider, as an example, a series of seven numbers (n), ordered and (ranked [R]): 3(1), 5(2), 7(3), 8(4), 11(5), 13(6), and 20(7). Estimate the 35th (p) percentile!

$$p = \frac{100 \times R}{n + 1} \tag{57}$$

and thus $R = \frac{p}{100} \times (n + 1) = \frac{35}{100} \times 8 = 2.8$. Accordingly, the 35th percentile corresponds to the rank R 2.8 and the numerical value corresponding to the rank shall now be calculated:

The p percentile is obtained by interpolation and is the value corresponding to

$Integer_R + Fraction_R \times$ (difference between value

scored R and $R + 1$)   (58)

In our example, $Integer_R$ is 2 and the corresponding value 5.

The next scored value is 7 and the $Fraction_R$ is 0.8. Thus,

$$p_{0.35} = 5 + 0.8 \times (7 - 5) = 6.16$$

This procedure will also give the correct median (50th percentile), i.e., $p = 0.50$: $R = \frac{50}{100} \times 8 = 4$ and the 4th number in the example is 8.

In an ordered data set, i.e., when the rank is known, the percentile is conveniently calculated from Equation (58), i.e., by interpolation.

The 25th percentile ($p(0.25)$) is called the lower quartile, and the 75th percentile ($p(0.75)$) is called the upper quartile.

Interquartile interval (IQR):

$$p(0.75) - p(0.25)$$ (59)

The interquartile interval includes 50 % of the data, which is also known as the central 50 % of the distribution. Since σ includes about 34 % of the distribution the interquartile interval, it corresponds to $0.5 \times 34/25 \times IQR$ or $0.68 \times IQR$.

Example

Calculate the interquartile interval using EXCEL: *PERCENTILE(interval,0.75) − PERCENTILE(interval,0.25)*.

Quantiles

Quantiles are points taken at regular intervals from the cumulative distribution function of a random variable. Dividing ordered data into q essentially equal-sized data subsets creates q-quantiles; the quantiles are the quantity values that mark the boundaries between consecutive subsets. Put another way, the kth q-quantile is the value x such that the probability that a random variable will be less than x is at most k/q and the probability that a random variable will be more than or equal to x is at least $(q-k)/q$.

There are $(q-1)$ quantiles, with k as an integer satisfying $0 < k < q$.

Yet another way to describe the quantile is as a number x_p such that a proportion p of the population values are less than or equal to x_p. The quantiles are thus values which divide the distribution so that there is a given proportion of the observations below the quantile. For example, the median is a quantile, the 50th percentile.

Thus, the 0.25 quantile (also referred to as the 25th percentile) of a variable is a value (x_p) such that 25 % (p) of the values of the variable fall below or are equal to that value.

Quantiles are used to investigate the normality of a distribution or to compare the type of frequency function between two distributions. The quantiles of an unknown distribution are calculated and compared with those of a known (normal) distribution in a regression analysis. If the distribution of the data coincides with the assumed distribution, the data will follow a linear regression line. This is recognized as a Q-Q (Quantile-Quantile) plot and gives an overall impression of one distribution in relation to another (Figure 5).

Quantiles are also used to estimate an empirical cumulative frequency plot, i.e., often called a mountain plot. The theory and estimation of mountain plots will be further discussed in the section on "Comparison of Methods."

More elaborate tests for normality are the *Kolgomorov-Smirnov* and *Anderson-Darling* tests which quantitate the divergence from normality.

Probit

The probit (probability unit) function is the inverse cumulative distribution function and the quantile function of the normal distribution. The probit function thus generates a value of a random variable, associated with a specified cumulative probability. The probit only exists for values above 0 and less than 1.

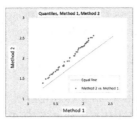 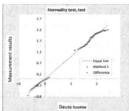 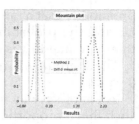

FIGURE 5 Quantile plots. In the left panel, the quantiles of two data sets are compared and the relatively linear regression indicates that the frequency distributions are similar but there is a bias. In the middle panel, the quantiles of one of the methods (high right) and those of the difference between the results (low left) are compared with a normal frequency distribution. Their close proximity to the equal line indicates that they are close to normal. In the right panel, the distribution of the data is shown as empirical cumulative plots "mountain plots" of the differences and one of the methods. At least the mountain plot of the differences is almost symmetrical, whereas the mountain plot of the results of method 1 shows a left skewness which is indicated also in the Q-Q plot. The vertical lines indicate the medians and delineate central 95 percentiles, respectively.

In EXCEL, the function is calculated by *NORMSINV (probability)* and can be used to simulate the sigmoid curve for probabilities between 0 and 1 (see Figure 3).

Logit

$$\text{logit}(p) = \ln\left(\frac{p}{1-p}\right) = \ln(\text{odds}) \tag{60}$$

where p is the probability.
The logit function is

$$\ln\left(\frac{p}{1-p}\right) = \ln(p) - \ln(1-p) = b + a \times X \tag{61}$$

The ratio between two odds is recognized as the odds ratio (R) (223). This is calculated as

$$R = \frac{\dfrac{p_1}{1-p_1}}{\dfrac{p_2}{1-p_2}} \tag{62}$$

Thus,

$$\ln(R) = \ln\left(\frac{\frac{p_1}{1-p_1}}{\frac{p_2}{1-p_2}}\right) = \ln\left(\frac{p_1}{1-p_1}\right) - \ln\left(\frac{p_2}{1-p_2}\right)$$
$$= \text{logit}(p_1) - \text{logit}(p_2) \tag{63}$$

Random Numbers

Random numbers can be obtained from tables or number generators (e.g., www.random.org) but are also available in EXCEL Thus, *RAND()* will return one single random number. *NORMSINV(RAND())* in an array will generate a set of normally distributed random numbers with an average of 0 and standard deviation of 1. The formula

$\$\bar{x} + \$s(x) \times NORMSINV(RAND())$

where $\bar{x}$ is the mean and $s(x)$ is the standard deviation, will modify the distribution of the random numbers accordingly.

An single value *NORMINV(RAND(),MEAN,SD)* copied to a set of cells will also generate a set of normally distributed values with the given mean and standard deviation. The program generates a new set of random number on all calculations, press F9.

Trimmed Means

To evaluate the effect of outliers (however defined), trimmed and winsorized[1] means may be used and are often referred to as robust location estimates. Trimmed and winsorized means should be used with care if the distribution is not symmetrical. The resulting distributions are not always Gaussian. On an average, trimming will underestimate the true dispersion.

Mean, trimmed:

$$\bar{x}_k = \frac{x_{(k+1)} + x_{(k+2)} + \cdots + x_{(n-k)}}{n - (2 \times k)} = \frac{1}{n - (2 \times k)} \sum_{i=k+1}^{n-k} x_i \tag{64}$$

---

[1] This procedure is named after the statistician Charles P. Winsor (1895-1951).

In the data set, the k lowest and k highest values are deleted. The arithmetic mean of the remaining data is the *trimmed mean*.

Example

See below.

Mean, winsorized: The winsorized mean resembles the trimmed mean, but rather than discarding the highest and the lowest numbers, the removed numbers are replaced with the next higher and next lower number, respectively:

$$w_k = \frac{(k+1) \times x_{(k+1)} + x_{(k+2)} + \cdots + x_{(n-k-1)} + (k+1) \times x_{(n-k)}}{n}$$

$$= \frac{k \times x_{(k+1)} + \sum_{i=k+2}^{n-k} x_i + k \times x_{(n-k)}}{n} \tag{65}$$

In a perfect Gaussian distribution, the trimmed and the winsorized means would remain unchanged but the standard deviation (dispersion) reduced.

Example

The results 802, 854, 823, 790, 815, 840, 833, 809, 843, 821 (mean 823.1; $s = 19.8$) are rearranged in increasing ordered: 790, 802, 809, 815, 821, 823, 833, 840, 843, 854.

To calculate the trimmed mean, delete 790 and 854:

$$\bar{x}_k = \frac{802 + 809 + 815 + 821 + 823 + 833 + 840 + 843}{8}$$

$$= 823.25; \quad s = 14.6$$

To calculate the winsorized mean, delete 790 and add 802 and delete 854 and add 843:

$$w_k = \frac{802 + 802 + 809 + 815 + 821 + 823 + 833 + 840 + 843 + 843}{10}$$

$$= 823.10; \quad s = 16.1.$$

Dispersion of Data

There are different measures of the dispersion, in addition to the standard deviation (variance); the median absolute deviation (MAD), the mean (average) absolute deviation (AAD), and the mean squared error (MSE) are the most often used. All are robust measures of dispersion and measures of the central tendency.

MAD,

$$\text{MAD} = \text{median}_i |x_i - \text{median}_d(x_d)| \tag{66}$$

i.e., the median of the observations' differences from their median.

Example

790, 802, 809, 815, 821, 823, 833, 840, 843, 854; median = 822.

Calculate the absolute difference from the median:

32, 20, 13, 7, 1, 1, 11, 18, 21, 32 and rearrange in ascending order:

1, 1, 7, 11, 13, 18, 21, 32, 32 and calculate the median of the differences from the median: MAD = (13 + 18)/2 = 15.5.

If

$$\frac{(\text{suspect number} - \text{median})}{\text{MAD}} > 5 \tag{67}$$

then the number is often regarded as an outlier.

Standard deviation based on MAD

$$\hat{s} \approx \frac{\text{MAD}}{0.6745} = 1.4826 \times \text{MAD} \tag{68}$$

provided the data are Gaussian distributed.

Multiple of Median, MOM:

$$\text{MOM} = \frac{x_i}{\text{median}} \tag{69}$$

MOM thus expresses the result as a fraction of the median and is a measure of how far an individual test result deviates from the median (cf. *z-score* (51)).

It is commonly used for instance in reporting the results of medical screening tests.

Mean absolute deviation (average deviation or mean absolute deviation)

$$\text{AAD} = \frac{\sum_{i=1}^{n}(|x_i - \bar{x}|)}{n} \tag{70}$$

- The AAD from the above example is 15.6.
- The EXCEL function $AVEDEV(x_1:x_n)$ calculates the AAD.
- The AAD must not be confused with MAD.
- The relation between these measures is $MAD < AAD < s$.

Further

$$\frac{AAD}{s} = \sqrt{\frac{2}{\pi}} = 0.7979 \tag{71}$$

provided the s is calculated from data that are normally distributed.

The MAD, AAD, and s all have the same unit as the original data.

Mean square error

$$MSE = \frac{\sum_{i=1}^{n}(x_i - x_0)^2}{n} \tag{72}$$

where x_0 is a true or a predetermined value according to a model. Compare RMS (27).

Uncertainty

The uncertainty concept is understood as an interval within which the true, or reference, value is supposed to be found with a defined level of confidence. Any bias is assumed to have been eliminated or corrected and substituted by an uncertainty that should be included in the uncertainty budget. This procedure to eliminate the bias will increase the uncertainty of the result but not necessarily change its value. An advantage with the uncertainty concept is that the laboratory takes the responsibility for containing a bias which the user normally cannot do.

New Chapter Measurement Uncertainty

The general formula for estimating error propagation is based on the partial derivatives (76) of the mathematical function (measurement function) to calculate the result. There are several shortcuts, and a frequently used method that is applicable to spreadsheet programs is that of Kragten (ref. see Eurachem Guide CG4 p. 104). The Kragten approximation requires that the variables are uncorrelated and it cannot handle complex formulas, e.g., exponentials. When the same quantity occurs more than once in a measurement function, the Kragten approximation may overestimate the uncertainty due to so-called compensating errors.

Propagation Rules

The combined uncertainty u_c of additions and subtractions

$$X = x_1 \pm x_2 \pm \cdots \pm x_i \tag{73}$$

is

$$u_c(X) = \pm \sqrt{u(x_1)^2 + u(x_2)^2 + \cdots + u(x_i)^2} \tag{74}$$

Example

The variables $A=10$ and $B=21$ have the standard uncertainties of $u(A)=0.3$ and $u(B)=0.6$; the sum a combined uncertainty $u_c(31) = \pm\sqrt{0.3^2 + 0.6^2} = 0.67$.

The combined uncertainty u_c of multiplications and divisions

$$X = x_i \times (\div)x_2 \times (\div) \cdots \times (\div)x_i \tag{75}$$

is

$$\frac{u_c(X)}{X} = \pm \sqrt{\left(\frac{u(x_1)}{x_1}\right)^2 + \left(\frac{u(x_2)}{x_2}\right)^2 + \cdots + \left(\frac{u(x_i)}{x_i}\right)^2} \tag{76}$$

Example

The ratio between the variables is $10/21=0.48$. Estimate the combined relative and absolute uncertainty of the ratio!

$$u_{c-rel} = \frac{u_c\left(\frac{A}{B}\right)}{\frac{A}{B}} = \frac{u_c\left(\frac{10}{21}\right)}{\frac{10}{21}} = \sqrt{\left(\frac{0.3}{10}\right)^2 + \left(\frac{0.6}{21}\right)^2} = 0.041$$

and the absolute uncertainty thus amounts to $u_c(0.48)=0.019$. If instead the numbers were multiplied, the relative uncertainty would be the same but the absolute different since the product is 210. Thus $u_c(210)=8.7$.

In many cases, the uncertainties or relative uncertainties, as appropriate, may be added linearly. This always leads to an overestimation of the combined uncertainty, the magnitude of which depends on the relation between the uncertainties of the terms and factors. In the examples above, the estimations would be 0.90 instead of 0.67 and 0.058 instead of 0.041, respectively.

A more complete method to estimate the uncertainty of a function $q(x, \ldots, z)$

$$u_c(q) = \sqrt{\left(\frac{\partial q}{\partial x} u(x)\right)^2 + \cdots + \left(\frac{\partial q}{\partial z} u(z)\right)^2} \tag{77}$$

In case the variables (e.g., $x, \ldots, z$) are not independent, then the covariance between all the variables must be taken into consideration

$$u_c(q)$$

$$= \sqrt{\left(\frac{\partial q}{\partial x} u(x)\right)^2 + \cdots + \left(\frac{\partial q}{\partial z} u(z)\right)^2 + 2 \times \frac{\partial q}{\partial x} \times \cdots \times \frac{\partial q}{\partial z} \times \text{covariance}} \tag{78}$$

For estimation and definition of the covariance, see Equations (177)–(179).

The result of an uncertainty analysis is summarized in an uncertainty budget, resulting in a combined uncertainty (u_c). The level of confidence may be adjusted by multiplying the combined uncertainty by a coverage factor k, to obtain the expanded uncertainty(U):

$$U(X) = k \times u_c(X) \tag{79}$$

The value of k shall always be attached to an expanded uncertainty. A k-value of 2 is generally accepted for a level of confidence of 95 %. The level of confidence does not have the same stringency as the "CI" which also considers the type of distribution (e.g., normal or Student's distribution).

Note The relative uncertainty is always calculated from the combined standard uncertainty, not the expanded uncertainty (cf. the Note to Equation 32).

Reference Change Value and Minimal Difference

As an example of error propagation, consider estimating the least significant difference between two results, in clinical chemistry known as "minimum difference," MD.

Example

If the results of two measurements are A and B with the uncertainties $u(A)$ and $u(B)$ a significant difference (D) must be larger than the uncertainty of the difference $u(D)$.

Therefore,

$$D > u(D) = \sqrt{u(A)^2 + u(B)^2}, \quad \text{cf.}(74) \tag{80}$$

Example

The desirable level of confidence is usually chosen to about 95 %, i.e., a k-value of 2 or 1.96 for a normally distributed data set. It may be reasonable to assume that $u(A)=u(B)$ and thus the MD is

$$MD > 1.96 \times u(A) \times \sqrt{2} = 2.77 \times u(A)$$

As a rule of thumb, the MD is often accepted as $MD = 3 \times u(A)$.

In clinical chemistry, the "reference change value," RCV, also includes the biological variation. The variation is usually expressed in relative terms ($\%CV(A)$):

$$RCV \geq \sqrt{2 \times \%CV(A_a)^2 + 2 \times \%CV(A_w)^2} =$$

$$\sqrt{\%CV(A_a)^2 + \%CV(A_w)^2} \times \sqrt{2} \tag{81}$$

where $\%CV(A_a)$ is the coefficient of variation of the measurement procedure and $\%CV(A_w)$ is the biological within individual variation.

Index of Individuality

The usefulness of reference values in diagnosis is often expressed as the index of individuality (II):

$$II = \frac{\sqrt{\%CV(A_w)^2 + \%CV(A)^2}}{\%CV(A_b)} \approx \frac{\%CV(A_w)}{\%CV(A_b)} \tag{82}$$

where $\%CV(A_b)$ is the between individual variation.

If low values ($II \leq 0.6$) are found the utility of reference values is usually limited in diagnosis. When $II \geq = 1.4$, the distribution of results for a single individual will cover a major part of the population reference interval and thus be of significant importance in diagnosis. A low II does not exclude the use of a quantity in monitoring a disease or condition in a specific individual.

Example

The II for S-Creatinine concentration is reported to about 0.3 and for S-Iron concentration about 1.1.

Type B Estimates of Uncertainty

The uncertainty of a measurement may be calculated by statistical means (Type A) or estimated by other methods, e.g., literature, experience (Type B). The rectangular and triangular distributions are frequently used in Type B estimates of the uncertainty. Estimates by Type A and Type B are treated equally in an uncertainty budget.

Standard uncertainty of a rectangular distribution:
If all results are distributed within an interval ($2a$) between an upper and a lower limit and the probability for a specific value is the same in the entire interval, then the distribution is known as "rectangular" or "uniform." This can also be expressed as that extreme values, close to the upper or lower limit, of the distribution will be as probable as anywhere within the distribution. No values are however expected or even possible outside the assumed interval.

The uncertainty estimated for a rectangular distribution is the most conservative, i.e., gives the largest standard uncertainty:

$$u(X) = \frac{a}{\sqrt{3}} \qquad (83)$$

where $2a$ is *Upper Limit – Lower Limit*.

Example

Consider the length of a rod placed far above you. Assume that it is no less than 50 cm and no longer than 150 cm and estimate the length and uncertainty! This means the best estimate is 100 cm, i.e., the middle of the interval and the standard uncertainty:

$$u(\text{rod}) = \frac{150 - 50}{2 \times \sqrt{3}} = \pm 28.9\,\text{cm}$$

Note An expanded uncertainty ($k=2$) will reach outside the assumed limits which always is the case since that will cover the probability that all observations are in the interval.

Standard uncertainty of a triangular distribution:
If a value is more likely than other values within an interval and no values expected or possible outside the interval, then a triangular distribution may be proper. Extreme values close to the limits of the assumed interval are possible but less likely than elsewhere in a symmetrical triangular distribution:

$$u(X) = \pm \frac{a}{\sqrt{6}} \tag{84}$$

The triangular distribution is attractive because of its simplicity. It is characterized by a lower limit and an upper limit (LL$=b$ and UL$=c$) and a mode d. The mean of the distribution is then

$$\mu = \frac{b + c + d}{3} \tag{85}$$

The variance is

$$\sigma^2 = \frac{(b - c)^2 + (b - d)^2 + (c - d)^2}{36} = \frac{b^2 + c^2 + d^2 - b \times c - b \times d - c \times d}{18} \tag{86}$$

In the case of a symmetrical triangular distribution, i.e., $d = \frac{b+c}{2}$, the square root of Equation (86) is equal to Equation (84), whereas in a "right-angle triangular distribution," i.e., $b=d$ it is simplified to

$$\sigma^2 = \frac{(b-c)^2}{18} = \frac{(2a)^2}{18} \tag{87}$$

The square root of the variance is the standard uncertainty:

$$u(X) = \frac{(b-c)}{3 \times \sqrt{2}} = \frac{2a}{3 \times \sqrt{2}}, \tag{88}$$

Since the expected value cannot be $<b$ or $>c$, the uncertainty is "one-sided," i.e., $b-u$ or $c+u$.

Standard uncertainty from an Gaussian distribution
When the measurement value is most likely to be near the center of an interval but there is a small, but real, possibility that there might be values outside the assumed or observed limits, then the appropriate "density function," i.e., distribution, is often assumed to be Gaussian. The standard uncertainty is estimated by

$$u(X) = \frac{a}{\sqrt{9}} \tag{89}$$

where $2 \times a$ is *Upper Limit − Lower Limit*.

Chi-Square (χ^2), an Index of Dispersion

Chi-square is the sum of the differences between found and expected number of observations squared, divided by the expected number:

$$\chi^2 = \frac{(f_1 - F_1)^2}{F_1} + \cdots + \frac{(f_n - F_n)^2}{F_n} = \sum_{i=1}^{n} \frac{(f_i - F_i)^2}{F_i} \tag{90}$$

where n is the number of "classes." f_i and F_i are sample counts of individuals (discrete quantities) which do and do not possess the property investigated and the corresponding hypothetical or expected frequencies being F_1 and F_2.

Chi-squared tests can only be used on actual numbers and not on percentages, proportions, means, etc. The *df* is the number of classes minus 1, i.e., $n-1$.

The expected number (F) is

$$F = n \times p; \quad p = \frac{F}{n} \tag{91}$$

Example

The probability of a particular event is 0.28. In an experiment of 35 observations, 16 events occurred and in 19 cases it did not. The expected occurrence is $0.28 \times 35 = 9.8$, and the not expected thus is $(1 - 0.28) \times 35 = 25.2$ cases. Is this what is expected?

$$\chi^2 = \sum_{i=1}^{n} \frac{(f_i - F_i)^2}{F_i} = \frac{(16 - 9.8)^2}{9.8} + \frac{(19 - 25.2)^2}{25.2} =$$
$$3.922 + 1.525 = 5.4$$

The df in this case is 1 and a table for chi-squared cumulative probabilities provides the critical χ^2-values 3.84 ($p = 0.05$) and 6.63 ($p = 0.01$).

The null hypothesis is rejected since the calculated χ^2 value is larger than the table value, i.e., there is a significant difference from what would be expected ($p < 0.05$).

This can be summarized in Table 5.

The sum of the (O-E) is always zero; accordingly if only two classes the $(O-E)^2$ will be equal for each class.

The EXCEL-function to evaluate χ^2 is $CHIINV(1 - \alpha, 1)$.

Thus, in this case $CHIINV((1 - 0.95), 1) = 3.84$.

The df for a contingency table equals (number of columns minus one) times (number of rows minus one) not counting the totals for rows or columns. For the 2×2 contingency table (see Tables 6 and 14), this gives $(2 - 1) \times (2 - 1) = 1$.

Example

The diagnostic sensitivity and specificity (see section on "Bayes' Theorem") of a particular quantity were 0.74 and 0.86, respectively. The prevalence of disease was 10 %. The performance of another biomarker was tested on the same

TABLE 5 Goodness of Fit

	Observed (O)	Expected (E)	$(O-E)$	$(O-E)^2$	$(O-E)^2/E$
	16	9.8	6.2	38.44	3.9
	19	25.2	−6.2	38.44	1.5
Total	35	35	0		5.4

TABLE 6 The Chi-Square is also Used in Evaluating Contingency Tables

Variable 2	Data Type 1	Data Type 2	Totals
Category 1	a	b	$a+b$
Category 2	c	d	$c+d$
Total	$a+c$	$b+d$	$a+b+c+d=N$

population, and the sensitivity and specificity were 0.70 and 0.77, respectively. Evaluate any difference between the performances.

The true positives (TP), the false negatives (FN), the true negatives (TN), and false negatives (FN) were calculated and the outcome summarized in two 2×2 tables.

	Quant A (Expected)			Quant B (Found)		
	"Positive"	"Negative"	Total	"Positive"	"Negative"	Total
Diseased	37	13	50	35	15	50
Non-diseased	63	387	450	105	345	450
Sum	100	400	500	140	360	500

Since the comparison was made using the same patient group, the number of diseased and nondiseased is the same in both trials. If the markers performed equally, the number in each cell would be the same and we can formulate the χ^2:

$$\chi^2 = \frac{(35-37)^2}{37} + \frac{(15-13)^2}{13} + \frac{(105-90)^2}{90} + \frac{(345-360)^2}{360} =$$
$$3.54; \quad df = 1;$$

The critical χ^2-value is 3.84 ($p=0.05$) *CHIINV(0.05,1)*, and therefore, the null hypothesis is accepted and it is not likely that there is a difference between the methods.

There is a short-cut method for estimating the χ^2-value from a 2×2 table:

$$\chi^2 = \frac{(a \times d - b \times c)^2 \times N}{(a+c) \times (a+b) \times (b+d) \times (c+d)} \tag{92}$$

Applying this formula to our example yields a χ^2-value of 3.47. The difference can be explained by rounding errors:

$p = 0.06$ *CHIDIST(3.47,1)*

The χ^2 test for small numbers of observations can be improved by Yates' continuity correction, which gives smaller values of χ^2:

$$\chi^2 = \sum_{i=1}^{n} \frac{(|f_i - F_i| - 0.5)^2}{F_i}; \quad df = 1 \tag{93}$$

Example A

$$\chi^2 = \frac{(35 - 37 - 0.5)^2}{37} + \frac{(15 - 13 - 0.5)^2}{13} + \frac{(105 - 90 - 0.5)^2}{90}$$

$$+ \frac{(|345 - 360| - 0.5)^2}{360} = 3.25; \quad df = 1$$

The corrected value is smaller than the uncorrected, but the difference is usually small enough not to change the conclusion. In practice, the Yates' correction has little influence on the outcome unless the total number of observations is less than 40.

Note that it is important how the expected value is calculated.

The χ^2 evaluation of a 2×2 table is an approximation, and the exact value can be calculated using Fisher's exact test, which is based on factorials:

$$p = \frac{\begin{pmatrix} a+b \\ a \end{pmatrix} \times \begin{pmatrix} c+d \\ c \end{pmatrix}}{\begin{pmatrix} N \\ a+c \end{pmatrix}} =$$

$$\frac{(a+b)! \times (c+d)! \times (a+c)! \times (b+d)!}{a! \times b! \times c! \times d! \times N!}$$

Example B

A classic example of the use of 2×2 tables is the evaluation of treating patients. Assume a study of treating hyperlipidemia with drug A and drug B. To estimate the expected number in each cell, it is assumed that there would be no difference between the treatments and the distribution between the cells equal to the distribution between the treated groups.

The expected number of hyperlipemic individuals treated by drug A (x):

$$\frac{x}{150} = \frac{240}{500}; \quad x = \frac{150 \times 240}{500}; \quad x = 72$$

Similarly, the expected number treated by drug B (y) would be

$$\frac{y}{150} = \frac{260}{500}; \quad y = \frac{150 \times 260}{500}; \quad y = 78$$

Those with no or less hyperlipidemia by the drugs A (z) and B (v):

$$\frac{z}{350} = \frac{240}{500}; \quad z = \frac{350 \times 240}{500}; \quad y = 168 \quad \text{and} \quad \frac{v}{350} = \frac{260}{500};$$
$$v = 182$$

	Hyperlipemia (Exp.)			Hyperlipemia (Obs.)		
	"Positive"	"Negative"	Total	"Positive"	"Negative"	Total
Drug A	72	168	240	110	130	240
Drug B	78	182	260	40	220	260
Sum	150	350	500	150	350	500

The rows and columns add up to the same number and the χ^2 calculated:

$$\chi^2 = \frac{(110 - 72)^2}{72} + \frac{(40 - 78)^2}{78} + \frac{(130 - 168)^2}{168} + \frac{(220 - 182)^2}{182} = 55.1; \quad df = 1$$

Since the χ^2 value is far above the critical value (3.84), the null hypothesis is discarded and the drug B more effective than drug A.

The contingency table can be expanded to having r rows and c columns:

$$\chi^2 = \sum_{i=1}^{n} \frac{(f_i - F_i)^2}{F_i}; \quad df = (r - 1) \times (c - 1) \tag{94}$$

$$F = \frac{\text{column total} \times \text{row total}}{\text{overall total}} \tag{95}$$

TABLE 7 Contingency Table for Several Categories

	Category I	Category II	Category III	Sum
Sample A	a	b	c	$a+b+c$
Sample B	d	e	f	$d+e+f$
Sample C	g	h	i	$g+h+i$
Sum	$a+d+g$	$b+e+h$	$c+f+i$	N

where f_i is the found and F_i the expected number in the corresponding cells, i.e., for cell a, the expected value would be $(a+b+c) \times (a+d+g)/N$; for cell b, $(a+b+c) \times (b+e+h)/N$; etc.

The results are inserted into Table 7 and the χ^2 calculated. The df would be $(3-1)\times(3-1)=4$. This expansion is not further discussed in the present text.

Chi-Square (χ^2), in Comparisons

The imprecision of a measurement method can be compared with the specification for the method:

$$\chi_c^2 = (n-1) \times \left(\frac{s}{s_0}\right)^2 \tag{96}$$

where s_0 is the specified (nominal) standard deviation and s the standard deviation found for the measurement procedure.

The calculated value χ_c^2 is compared to the table value at the appropriate degrees of freedom:

$$\chi_{\text{crit}}^2 = \frac{\chi_{c(\alpha;(n-1))}^2}{n-1} \tag{97}$$

where n is the number of observations and $(n-1)$ is the degrees of freedom. If $\chi_c^2 \geq \chi_{\text{crit}}^2$, the null hypothesis $h0$: $s \leq s_0$ is rejected.

Example

In a verification procedure, the standard deviation of 10 repeated measurements was 0.25 mmol/L. The manufacturer claimed an uncertainty of 0.20 mmol/L. Is the found standard deviation reasonable at a level of confidence or $\alpha = 0.05$?

$$\chi_c^2 = (n-1) \times \left(\frac{s}{s_0}\right)^2$$

$$= (9) \times \left(\frac{0.25}{0.20}\right)^2 = 14.1; \quad \text{the } \chi^2_{crit(0,05,9)} = 16.9$$

Since the calculated value does not exceed the critical table value, the claim is not rejected. If, however, the standard deviation was based on 30 repeated measurements, the χ_c^2 and $\chi^2_{crit(0,05,14)}$ equal 45.3 and 42.6, respectively, and the claim would be rejected. The rational is that the standard deviation would have been estimated with a considerably smaller CI.

In EXCEL, the $\chi^2_{c(\alpha;(n-1))}$ is calculated as *CHIINV(α,df)*.

The Rule of Three

This rule states that if no event happens in n observations, then it can be assumed, with a probability of 95 %, that it will happen less frequently than 1 in $n/3$.

This rule is an acceptable approximation if $n > 30$.

Example

If in a series of measurements there is no outlier in 600 consecutive measurements, it can be concluded with a 95 % confidence that there will be less than 1 outlier in 200 measurements, i.e., less than 0.5 %.

ANALYSIS OF VARIANCE

Definitions and Calculation

The one-way ANOVA was originally designed to allow comparison of several means of data sets rather than using a series of Student's independent t-tests (123) for all possible pairs. The procedure will evaluate if there is a difference between the means of the studied groups. The ANOVA calculates the sum of squares within the groups and between the groups. The significance of a difference is established using F-statistics (132). Solution of an ANOVA experiment is offered

TABLE 8 Standard ANOVA Output Table

Between groups	SS_b	df_b	MS_b	$F = MS_b/MS_w$	p-Value
Within groups	SS_w	df_w	MS_w		
Total	SS_{tot}	df_t			

SS is the "sum of squares"; df, degrees of freedom; and MS, "mean square." The F-value indicates the significance of the difference between groups in the study.

TABLE 9 Notations Used in Describing an ANOVA

	Group 1	Group 2	. . .	Group k
Sample 1	x_{11}	x_{21}	. . .	x_{k1}
Sample 2	x_{12}	x_{22}	. . .	x_{k2}
.	.	.	.	.
.	.	.	.	.
Sample n	x_{1n}	x_{2n}	. . .	x_{kn}

by all standard statistical packages and built into many spreadsheet programs. EXCEL supports one-way ANOVA, but this requires that the "Data analysis Add-in" has been installed. It is a part of the standard program but needs activation. It will then be found under "Data." The solution is often presented in a standardized format (Table 8).

The experimental design comprises several (k) groups, runs, or series of results each including several (n_k) observations (x), with a total number of $N = kxn_k$ observations. Groups may comprise different numbers of observations (unbalanced design). The group means are designated $\bar{x}_k$ and the grand mean $\bar{\bar{x}}$ (Table 9).

The procedure is to calculate the sum of squares (SS) for the total, within and between groups.

Total:

$$SS_{tot} = \sum_{i=1}^{i=N} (x_i - \bar{\bar{x}})^2 = \sum_{i=1}^{i=N} x_i^2 - \frac{\left(\sum_{i=1}^{i=N} x_i \right)^2}{N} \tag{98}$$

$$df_{tot} = N - 1 \tag{99}$$

Alternative calculation:

$$SS_{tot} = (N-1) \times Var(x_i) \tag{100}$$

Between groups:

$$SS_b = n_0 \times \sum_{i=1}^{i=k} (\bar{x}_i - \bar{\bar{x}})^2 \tag{101}$$

where n_0 is the number of observations in the groups. If this varies between groups (an unbalanced design), then Equation (101) is rearranged to

$$SS_b = n_0 \times \sum_{i=1}^{i=k}(\bar{x}_i - \bar{\bar{x}})^2 = \sum_{i=1}^{i=k}\left(n_i \times (\bar{x}_i - \bar{\bar{x}})^2\right) =$$

$$\sum_{i=1}^{i=k} n_i \times \bar{x}_i^2 - \frac{\left(\sum_{i=1}^{i=N} x_i\right)^2}{N}; \tag{102}$$

$$df_b = k - 1 \tag{103}$$

If the design is unbalanced, use a "weighted" grand mean

$$\bar{\bar{x}} = \frac{\sum_{i=1}^{i=k} n_i \times \bar{x}_i}{\sum_{i=1}^{i=k} n_i} = \frac{\sum_{i=1}^{i=k} n_i \times \bar{x}_i}{N} \tag{104}$$

Within groups:

$$SS_w = (N - k) \times \frac{(n_1 - 1) \times s_1^2 + \cdots + (n_k - 1) \times s_k^2}{n_1 + \cdots + n_k - k} =$$

$$\sum_{i=1}^{i=k}\left((n_i - 1) \times s_i^2\right) \tag{105}$$

Compare the calculation of SS_w with the that of the pooled standard deviation (33)!

If the groups comprise the same number of observations (balanced), then

$$SS_w = \sum_{i=1}^{i=k}\left((n_i - 1) \times s_i^2\right) = \sum_{i=1}^{n_i}(x_i^2) - \sum_{i=1}^{n_i} n_i \times \bar{x}_i^2$$

$$df_w = N - k \tag{106}$$

$$SS_{tot} = SS_b + SS_w \tag{107}$$

Thus, it may be practical to use the simple formula for SS_{tot} (100) and subtract either SS_w or SS_b to calculate the SS_b and SS_w, respectively.

ANOVA to Evaluate Differences Between Means

To evaluate if there is a difference between the means of several groups, an F-test is performed. In the ANOVA evaluation, the F is calculated as

$$F = \frac{MS_b}{MS_w} \tag{108}$$

The F is evaluated using an F-table.
In EXCEL, use $FINV(prob,df1,df2)$

Note In estimating the F-value, the MS_b is always in the numerator and MS_w in the denominator. See also Equations (131) and (132).

A high F-value will indicate a significant difference between groups but not between *which* groups. There are different techniques to estimate the significance level of differences between groups.

A simple approach is to arrange the means in increasing or decreasing order and calculate the "least significant difference":

$$\Delta_{sign} = t_{(1-\alpha),(n-2)} \times s_{within} \times \sqrt{\frac{2}{n}}, \tag{109}$$

where s_{within} is the estimated within group standard deviation $\left(\sqrt{MS_{within}}\right)$, n is the number of observations in each group (balanced design), $t(n-2)$ the t-value for the indicated degrees of freedom.

Comparison of this value with the difference between the ordered means will indicate where a significant difference may be found. This formula is derived from Equation (125) (independent Student's t), assuming that the standard deviation is the same for the groups and the difference between the means is the difference that is tested for significance. This is directly seen by rearranging the formula to

$$t_{(1-\alpha),(n-2)} = \frac{\Delta_{sign}}{\sqrt{\frac{2 \times s_{within}^2}{n}}} \tag{110}$$

A formula applicable in unbalance designs with different standard deviations would then be

$$\Delta = t_{(1-\alpha),\,(n-2)} \times \sqrt{\frac{s_1^2}{n_1} + \frac{s_2^2}{n_2}} \quad \text{and} \tag{111}$$

Remember though that the standard deviation of the groups needs to be similar and the degrees of freedom may need to be calculated using the Satterthwaite's approach (see Equation 128).

There are more sophisticated and rigorous solutions to this problem. It is advised not to use the Student's t-test repeatedly to avoid the risk false significances.

Example

The S-Cholesterol concentration was measured in 4 groups of 10 participants on different diets:

	1	2	3	4	5	6	7	8	9	10	Mean	Var
1	5.2	5.6	6.8	3.5	5.9	6.3	7.2	8.5	6.8	5.7	6.15	1.77
2	6.2	6.2	7.9	8.2	5.7	6.6	8.1	9	12.1	6.7	7.67	3.57
3	4.2	4.9	6.8	3.7	4.5	5.6	6.4	5.8	6	7	5.49	1.26
4	4.8	7.1	5.9	4.7	5.8	4.9	5.6	7.5	5.1	5.8	5.72	0.89

"Grand mean": 6.26; total variance: 2.47:
SS_{tot}: $(40 - 1) \times 2.47 = 96.50$; $df = 40 - 1 = 39$.
SS_b: $10 \times (6.15 - 6.26)^2 + (7.67 - 6.26)^2 + (5.49 - 6.26)^2$
$+(5.72 - 6.26)^2 = 28.85$; $df = 4 - 1 = 3$.
$SS_w = (10 - 1) \times (1.77 + 3.57 + 1.26 + 0.89) = 67.65$;
$df = 40 - 1 - 3 = 36$.
$$F = \frac{SS_b}{SS_w} \times \frac{df_w}{df_b} = 5.12$$

The result indicates that there is significance between the groups.

It is convenient but not necessary to sort the calculated means in ascending order: 5.49, 5.72, 6.15, and 7.67.

Since clearly the variances (standard deviations) of the groups are different formula (111) is applicable. Reviewing groups 2 and 4:

$$\Delta_{sign} = t_{0.05,\,18} \times \sqrt{\frac{3.57}{10} + \frac{0.89}{10}} = 2.10 \times 0.668 = 1.40$$

A cross-table to display the differences between the means:

	5.49	5.72	6.15	7.67
5.49	0	0.23	0.66	2.18
5.72		0	0.43	1.95
6.15			0	1.52
7.67				0

The number of unique entries in a cross-table is $\frac{n \times (n+1)}{2}$, where n is the number of observations.

Thus, there are significant differences between the highest (group 2) and the remaining three groups between which there is no significant difference.

The data set analyzed by the ANOVA procedure in EXCEL generates the table:

	SS	df	MS	F	p-Value	F-Crit
Between groups	28.85	3	9.62	5.12	0.00	2.87
Within groups	67.65	36	1.88			
Total	96.50	39				

Nonparametric Methods

The use of ANOVA assumes that the data are normally distributed and that the variances of within the groups are of a similar magnitude (cf. Student's t-test) within the measuring interval.

The *Kruskal-Wallis test* is a nonparametric alternative to the one-way ANOVA, and the *Friedman's test* can be compared with the two-way ANOVA. In both tests, all the observations are ranked together, any ties given the same calculated rank. Then sum of the ranks in each method is used to calculate statistics that can be evaluated by comparing with a χ^2 table. Both procedures can be applied to ordinal, interval, and rational data. The test will only demonstrate that there is at least one group which differs from the rest. Posttests may be required to identify significant differences.

Analysis of Variance Components

If the same sample is analyzed repeatedly in several series, the ANOVA can be used to estimate the within- and between-series variance and from them the combined variance. Thus, the mean squares (MS) of the ANOVA table are equivalent to the between- and within variance (s^2) obtained by dividing the sum of squares by the corresponding degrees of freedom:

$$MS_b = \frac{SS_b}{df_b} \tag{112}$$

$$MS_w = \frac{SS_w}{df_w} \tag{113}$$

The *within-series variance* (MS_w) is equivalent to the pooled variance (33) of the runs times the degrees of freedom $(N - k)$.

The *between-series variance* can also be calculated directly as the average number of observations (n_0) in each group times the variance of the group means s_g^2:

$$s_g^2 = \frac{\sum_{i=1}^{i=k}(\bar{x}_i - \bar{\bar{x}})^2}{(k-1)} \tag{114}$$

$$MS_b = n_0 \times \frac{\sum_{i=1}^{i=k}(\bar{x}_i - \bar{\bar{x}})^2}{(k-1)} = n_0 \times s_g^2 \tag{115}$$

The total (combined) variance can only be estimated after compensation for the contribution from the within-series variance of the MS_b:

"Purified" (or "pure") between run and intermediary precision:

$$s_b = \sqrt{\frac{MS_b - MS_w}{n_0}} = \sqrt{s_g^2 - \frac{s_{pool}^2}{n_0}} \tag{116}$$

where n_0 is the average number of observations in each group, run, or series.

However, if the number of observations in the groups differs, then a "harmonic" average number of observations should be used:

$$n_0 = \frac{N^2 - \sum_{i=1}^{i=k}(n_i)^2}{N \times (k-1)} = \bar{n}_i - \frac{s(n_i)^2}{N} \tag{117}$$

where N is the total number of observations and n_i is the number of observations in each group and k is the number of groups. In most cases, the difference between the arithmetic mean of the number of observations and Equation (117) is negligible (the second term in the second form of (117)) in practical work.

s_b^2 is also known as the "unbiased estimate of the between group variance."

Combined uncertainty:

$$s_{tot} = u_c(x) = \sqrt{(s_b)^2 + MS_w} \tag{118}$$

As seen from Equation (116), the MS_b cannot be less than the MS_w, which would require the square root of a negative number. If, however, $MS_b < MS_w$, then the s_{tot}, by convention, is set to $\sqrt{MS_w}$, i.e., $MS_b = 0$. This condition can be formulated as

$$s_b^2 = MAX\left(0, \frac{MS_b - MS_w}{n_0}\right) \tag{119}$$

which is also how it is coded in EXCEL.

The total variance of several series of values can be calculated by different approaches. Consider for instance, the total estimated variance of results from a laboratory with several instruments performing the same measurements. The s_{tot} could be the estimated variance from the total data set ($\sqrt{VAR(x_{11} : x_{kn})}$, where k represents the number of series and n the number of observations in the series thus $x_{11}:x_{kn}$ reads from the first to the last). It can be argued that a representative variance is the average of the variance of the series. As above, the s_{tot} can be estimated by the ANOVA components. The difference between these approaches is minimal if there is none or a very small (less than about 1 %) between-series difference. In all other cases, the averaged variance underestimates the s_{tot} and that estimated from the total data set either over- or underestimates the s_{tot},

TABLE 10 Results of Repeated Measurements

	Series 1	Series 2	Series 3	Series 4	Series 5
Result 1	124	125	120	118	125
Result 2	125	122	122	120	126
Result 3	122	127	123	120	128
Result 4	126	125	122	118	124
Result 5	124	122	123	120	124
Mean	124.2	124.2	122.0	119.2	125.4
Standard deviation	1.5	2.2	1.2	1.1	1.7

TABLE 11 Standard ANOVA Table Normally Displayed

Source of Variation	SS	df	MS	F	p-Value
Between groups	120.4	4	30.1	12.1	<0.001
Within groups	49.6	20	2.48		
Total	170	24			

the ANOVA components taken as the gold method. The within-series variance plays a minor role.

Example

Control material was measured five times in five series (Table 10 and 11). Evaluate the difference between the means of the series and calculate the within-, between-, and combined uncertainties!

$$s_b^2 = \frac{(30.1 - 2.48)}{5} = 5.52; \quad s_b = 2.4$$

$$s_w^2 = 2.48; \quad s_w = 1.5$$
$$u_c(x) = \sqrt{2.48 + 5.52} = \sqrt{8.00} = 2.8$$

$$\sqrt{VAR(x_{11} : x_{kn})} = \sqrt{7.08}; \quad s(x) = STDEV(x_{11} : x_{kn}) = 2.7$$

Conclusions: there is a significant difference between the series ($F = 12.1$). The major source of the combined uncertainty is the between-series variation.

Youden Plot

The Youden plot is a graphical method to estimate and visualize random and systematic errors in measuring identical or similar samples of two different concentrations or from two or more sites (laboratories or instruments). The original Youden plot was designed to identify the random and systematic errors in many laboratories, e.g., in evaluating EQA or PT (External Quality Assessment, Proficiency Testing) experiments. In these schemes, the same samples of two or more concentrations are distributed simultaneously over time, to several participating laboratories.

The principle of the Youden plot can also be used in one laboratory where several samples are measured in duplicates or by different measurement procedures.

Samples are measured either in duplicates (X_i and Y_i, where i identifies the sample) or the same samples in different laboratories or by different procedures (where i then identifies the laboratory or procedure).

The results are plotted (Y_i vs. X_i) in a two-dimensional scattergram with the same scale of the axes. Horizontal and vertical lines through the median (Manhattan median) of the results creates four quadrants which are crossed by a diagonal (slope 1) $Y = X + a$, through the median (Figure 6).

Results with agreeing results will be found in the quadrants of the diagonal and thus correspond to the "true-positive" and "true-negative" results as discussed in the section on "Bayes' Theorem." Results in the remaining quadrants represent the "false-negative" and the "false-positive" results.

A systematic error will make the X_i/Y_i move away from the median along the diagonal, and random errors will move the coordinates away from the diagonal into the "false quadrants." The distribution of the points in the four quadrants will give a fair overview of the random and systematic errors encountered. To assist the evaluation, a circle or a rectangle/quadrate is often displayed around the median (Figure 6).

Results from input to a Youden plot can be used for variance analysis as a simplified alternative to a two-way ANOVA. If the same or similar samples are measured using two procedures (or in two laboratories), their difference can reasonably

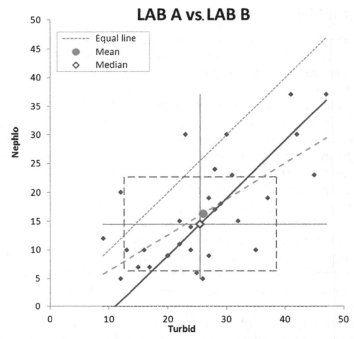

FIGURE 6 Youden plot. Solid vertical and horizontal lines delineate quadrants with false-negative, true-positive, false-positive, and true-negative relations between results. Solid 45° line is the diagonal, the dotted line, the equal line and the hatched line the regression line. The rectangle illustrates the acceptable random and systematic deviations.

be assumed to cancel a systematic error. $D = (X_i + \varepsilon) - (Y_i + \varepsilon)$. The distribution of the differences will therefore be an estimate of the repeatability (r) standard deviation s_r.

$$(s_r)^2 = \frac{\sum_{i=1}^{n}(D_i - \bar{D})^2}{2 \times (n-1)} = \frac{\mathrm{Var}(D_i)}{2} \tag{120}$$

The sum of the results (S) gives an estimate of the overall variation or reproducibility (R). The overall variance $(s_R)^2$ is

$$(s_R)^2 = \frac{\sum_{i=1}^{n}(S_i - \bar{S})^2}{2 \times (n-1)} = \frac{\mathrm{Var}(S_i)}{2} \tag{121}$$

The spread due to interindividual variance, $(s_L)^2$, will then be

TABLE 12   The Concentration of 10 Samples, Measured with 2 Instruments

	1	2	3	4	5	6	7	8	9	10	Mean	S
1	5.2	5.6	6.8	3.5	5.9	6.3	7.2	8.0	6.8	5.7	6.10	1.24
2	4.8	7.1	6.3	4.7	5.8	4.9	5.6	7.5	5.1	5.8	5.76	0.96

Overall mean: 5.93.

$$(s_R)^2 = 2 \times (s_L)^2 + (s_r)^2; \quad (s_L)^2 = \frac{(s_R)^2 - (s_r)^2}{2} \tag{122}$$

A factor 2 in the denominator is necessary because D and S are estimated in two sets of results.

These calculations should not be confused with the Dahlberg approach (36) to estimate the standard deviation of a set of duplicate measurements of *different* samples (Table 12).

Example

$$(s_r)^2 = \frac{\text{Var}(D_i)}{2} = 0.59; \quad s_r = 0.77$$

$$(s_R)^2 = \frac{\text{Var}(S_i)}{2} = 0.67; \quad s_R = 0.82$$

$$(s_R)^2 = 2 \times (s_L)^2 + (s_r)^2; \quad (s_L)^2 = \frac{(s_R)^2 - (s_r)^2}{2} = \frac{0.67 - 0.59}{2}$$
$$= 0.04, \quad s_L = 0.20$$

The relative overall (repeatability) standard deviation: $\%CV = \frac{100 \times 0.82}{5.93} = 13.79$.

DIFFERENCE BETWEEN RESULTS; STUDENT'S t-TESTS

Difference—Two Scenarios

Two different scenarios can be identified in the estimation of the significance of a difference between results:

(1) the difference between the means of two different data sets (e.g., the difference in the concentration of an analyte in men and women) and

(2) the difference between paired results (e.g., the individual effect of a treatment).

The difference between the estimated mean and the true value μ is expressed in relation to the standard error of the mean (SEM):

$$|t| = \frac{\bar{x} - \mu}{\frac{s_x}{\sqrt{n}}} = \frac{(\bar{x} - \mu) \times \sqrt{n}}{s_x} \tag{123}$$

Note the similarity to the z-score (51) in which a difference is expressed in standard deviations rather than SEM.

Difference Between Paired Results; Students t_{dep}

If a quantity is measured in the same sample or individual before and after an intervention and n pairs measured, and the difference between the results is d_i, then

$$|t_{dep}| = \frac{\bar{d}}{\frac{s_d}{\sqrt{n}}} = \frac{\frac{\sum_{i=1}^{i=n} d_i}{n}}{\frac{s_d}{\sqrt{n}}} = \frac{\frac{\sum_{i=1}^{i=n} d_i}{n} \sqrt{n}}{s_d} = \frac{\sum_{i=1}^{i=n} d_i}{s_d \times \sqrt{n}} \tag{124}$$

where $\bar{d}$ is the average of the differences between the pairs and s_d its standard deviation.

Degrees of freedom for t_{dep}:

$$df = n - 1 \tag{125}$$

EXCEL supports the t_{dep}; the routine requires that an "Add-in" is installed. It will then be found under "Data" and is called "paired two sample for mean".

Since the Student's t_{dep}-test is based on an estimated mean and standard deviation, it assumes a Gaussian distribution of the differences between the observations. The distribution of the data set itself is not important. If the difference is not Gaussian distributed, the nonparametric method of choice is *Wilcoxon signed-rank test* (see below).

A χ^2-test (chi-square), "sign test," with or without correction for continuity for small samples may also be used.

Difference Between Means; Student's t_{ind}:

A comparison between two means, when the individual observations are independent, is known as Student's t_{ind}.

The t-value as defined in Equation (123) is also known as the single sample t-test since it assumes a comparison with μ which is assumed to have no variance. The degrees of freedom in the single sample t-test will be $n_1 - 1$.

If, however, the comparison is between two means, $\bar{x}_1$ and $\bar{x}_2$, with the standard deviations s_1 and s_2, and n_1 and n_2 observations in the groups, respectively, the t-value is estimated according to Welch:

$$|t_{\text{ind}}| = \frac{\bar{x}_1 - \bar{x}_2}{\sqrt{\dfrac{s_1^2}{n_1} + \dfrac{s_2^2}{n_2}}} = \frac{\bar{x}_1 - \bar{x}_2}{\sqrt{s(\bar{x}_1)^2 + s(\bar{x}_2)^2}} \tag{126}$$

where $s(\bar{x})$ is the standard error of the mean, also abbreviated SEM.

If $s_1 \sim s_2$ is the degrees of freedom for t_{ind}:

$$df = n_1 + n_2 - 2 \tag{127}$$

and accordingly

$$t_{\text{ind}} = \frac{\bar{x}_1 - \bar{x}_2}{s \times \sqrt{\dfrac{1}{n_1} + \dfrac{1}{n_2}}} \tag{128}$$

If $s_1 \neq s_2$, however, use the Welch-Satterthwaite approximation to estimate the df:

$$df = \frac{\left[\left(\dfrac{s_1^2}{n_1}\right)+\left(\dfrac{s_2^2}{n_2}\right)\right]^2}{\dfrac{\left(\dfrac{s_1^2}{n_1}\right)^2}{n_1-1}+\dfrac{\left(\dfrac{s_2^2}{n_2}\right)^2}{n_2-1}} = \frac{\left[\left(\dfrac{s_1^2}{n_1}\right)+\left(\dfrac{s_2^2}{n_2}\right)\right]^2}{\dfrac{s_1^4}{n_1^2(n_1-1)}+\dfrac{s_2^4}{n_2^2(n_2-1)}} = \frac{\left(se_1^2+se_2^2\right)^2}{\dfrac{se_1^4}{(n_1-1)}+\dfrac{se_2^4}{(n_2-1)}} \tag{129}$$

EXCEL supports two procedures for calculating the t_{ind}: one when equal variances are assumed "Two Sample Assuming Equal Variances" and "Two Sample Assuming Unequal Variances."

The significance of a difference between variances (standard deviations) and thus the need to apply the Welch-Satterthwaite approximation can be estimated applying an F-test (131).

Note The *df* according to this formula does not necessarily yield an integer and may be rounded up, or down, to the nearest integer in evaluating the obtained *t*-value from a table. The use of the Welch-Satterthwaite approximation gives a conservative estimate of the significance.

The uncertainty of the quantity value of a reference material is usually expressed as a CI or as the standard error, i.e., how well the value has been determined. Assuming a SEM of the reference material to be $s(\bar{x}_{RM})$ the formula will become

$$t_{ind} = \frac{\bar{x}_1 - \bar{x}_{RM}}{\sqrt{\frac{s_1^2}{n_1} + s(\bar{x}_{RM})^2}} \tag{130}$$

The quantity value of the reference material may be an assigned value without uncertainty attached. Then Equation (130) can be further simplified.

The Student's independent *t*-test assumes a Gaussian distribution of the data sets. If this is not the case, the nonparametric method of choice is *Mann-Whitney U-test* which is described in some detail below. An alternative to identify a difference between samples is the *Tukey's quick test*.

A Comparison Between Many Series

If many series are compared with successive Student's *t*-test and they are independent, then the overall probability $(1 - \alpha)$ to conclude that there is a difference although there is none, will increase. This is because the probability when we test the null hypothesis of the comparisons is the product of the individual probabilities $(1 - \alpha_1) \times (1 - \alpha_2) \times (1 - \alpha_3)\ldots$

If α is 0.05, then for three series the overall probability would be $0.95 \times 0.95 \times 0.95 = 0.857$. The probability that at least one error occurs is $1 - 0.857 = 0.143$ which explains and quantifies the risk to assume a difference when there is none. A simple correction, namely, the Bonferroni correction, is to divide the α (in this case 0.05) with the number of comparisons, i.e., 3 and the combined α equals 0.017. This is the table value for each of the comparisons, and the degrees of freedom (*df*) is $n - 2$ as used for the *t*-test; EXCEL:T.INV.2T(α;*df*) for a two-sided comparison and $TINV(1-\alpha;df)$ for a one-sided comparison.

In essence, the Bonferroni correction makes it more difficult to achieve and eventually exceed the critical *t*-value.

The Bonferroni correction is regarded as a conservative test and in other words may make it "more difficult" to demonstrate significance. This leads to a decreased statistical power which may be overcome by including more observations in the comparisons.

Interpretation of a *t*-Value

Estimated *t*-values are interpreted in a *t*-table. The entries to this table are the *df* and the probability. For a given *df* find the value below, and as close to the estimated value as possible in the table. The column in which it is found represents the probability. Thus, an estimated *t*-value can lead to different interpretations depending on the number of observations. The higher the *t*-value, the less probable it is that the compared quantities are the same (the null hypothesis true).

Example

A *t*-value of 2.2 was obtained in a study. The *df* was 20 (Table 13).

Thus, the probability (*p*-value) that the null hypothesis was true was <0.05. Note that the interpretation of the *t*-value is independent of how the *t*-value was estimated. Using the EXCEL $TINV(0.05,20)=2.086$.

TABLE 13 Extract of a *t*-Table

df	Probability of Two-Tailed Test			
	0.1	0.05	0.01	0.001
9	1.83	2.26	3.25	4.78
18	1.17	2.10	2.88	3.92
20	1.17	2.09	2.85	3.85

Example

Suppose we have the data set:

	1	2	3	4	5	6	7	8	9	10	Mean	$s(x)$
1	5.2	5.6	6.8	3.5	5.9	6.3	7.2	8.0	6.8	5.7	6.10	1.24
2	4.8	7.1	6.3	4.7	5.8	4.9	5.6	7.5	5.1	5.8	5.76	0.96
Diff.	0.4	−1.5	0.5	−1.2	0.1	1.4	1.6	0.5	1.7	−0.1	0.34	1.09

Assume further that the data have been collected from a normally distributed data set and further that the difference between the observations is also normally distributed and that the variances of the sample results are not significantly different (see below). We can use the data in two examples.

One is to assume that rows 1 and 2 are from different experiments and the task is to investigate if the means are different. The second is to look upon the first row as results obtained before a treatment and the second the results after the treatment.

In the first case, we apply the t_{ind} and in the second the t_{dep}:

1. $|t_{ind}| = \dfrac{\bar{x}_1 - \bar{x}_2}{\sqrt{\dfrac{s_1^2}{n_1} + \dfrac{s_1^2}{n_1}}} = \dfrac{6.10 - 5.76}{\sqrt{\dfrac{1.24^2 + 0.96^2}{10}}} = \dfrac{0.34}{\sqrt{0.25}} =$

$\dfrac{0.34}{0.50} = 0.68; \quad df = 10 + 10 - 2 = 18$

From a t-table, we find that the t-value for $df = 18$ should exceed 2.1 to be significant in a two-sided test.

2. $|t_{dep}| = \dfrac{\bar{d}}{\dfrac{s_d}{\sqrt{n}}} = \dfrac{0.34 \times \sqrt{10}}{1.09} = 0.99; \quad df = 10 - 1 = 9$

From a t-table, we find that the t-value for $df = 9$ should exceed 2.3 to be significant in a two-sided test. This information can also be retrieved by the *TINV(Probability, df)* function in EXCEL.

Comparison of Variances

To answer the question if there is a significant difference between the variances of the results of two samples, the F-test may be used:

$$F = \frac{s_1^2}{s_2^2} \tag{131}$$

The larger s is always in the numerator; thus the F-value is always ≥ 1. The F-value is interpreted using an F-table. The F-table considers the number of observations in both groups, i.e., the df may be different in the samples $(n_1 - 1)$ and $(n_2 - 1)$, respectively.

For a one-sided test (95 % probability), use $\alpha = 0.05$, for a two-sided test $\alpha = 0.05/2$.

The critical F-value is calculated in EXCEL as $FINV(\alpha, (n_1 - 1),(n_2 - 1))$.

Note The EXCEL requires that the number of observations with the larger variance should be entered first. EXCEL carries only out a one-sided test and the α-value should be chosen accordingly if a two-sided test is desired.

Example

The variances in the example from the t-test were 1.54 and 0.93. The F-value is 1.66. In EXCEL, $FINV(0.05,9,9) = 3.18$ for a one-sided test and thus the variances are not significantly different since the critical value (3.18) is not exceeded.

The F-test may also be used to answer the question if a method (*new*) is significantly more precise than another (*old*), this is a one-tailed use:

$$F_c = \frac{s_{old}^2}{s_{new}^2} \tag{132}$$

In either case, the variances are considered significantly different if $F_c > F_{crit}$ (table value or $FINV(\alpha,(n_1 - 1),(n_2 - 1))$).

Since the F-*test* is based on variances, it assumes a Gaussian distribution of the data.

NONPARAMETRIC COMPARISONS

In general, conclusions drawn from nonparametric methods are less powerful than parametric when the distribution is known and its properties can be applied. However, as nonparametric methods make fewer assumptions, they are more flexible, more robust, and sometimes applicable to ordinal data.

The nonparametric test for dependent data, corresponding to the *Student* t_{dep}, is the Wilcoxon signed-rank test. To compare independent data, the Mann-Whitney's U-test corresponds to that of Student's t_{ind}.

Wilcoxon Sign Rank Test for Paired Samples

The significance of a difference between paired observations can be evaluated by a nonparametric test, usually recognized as the *Wilcoxon sign rank test*. This is used to test the null hypothesis that there is no difference between the paired observations or in other word if the ranked pairs in the positive and negative groups differ from what would be expected. The basic assumption is thus to estimate if the probability that the obtained number of positive (W^+) and negative (W^-) differences is what would be expected in view of how many observations have been made and applying the binominal theorem. If the number of observations is large (e.g., >20), the distribution may be normalized and the Wilcoxon test solved by coding in EXCEL as described below.

Strictly, a condition for applying this test is that the distribution is symmetrical.

The differences of the paired observations are sorted and ranked, disregarding the sign of the difference, i.e., the absolute numbers of the results are ranked. Thus, the differences have been exchanged for their ranks in an ordered data set. In EXCEL, the ranking can be achieved without physically sorting the data by using the *RANK* function, e.g., *RANK(ABS(A2),$A$2:$D$10)*, where $A$2:$D$10 is the data set and ABS(A2) is the absolute value of the first difference in the data set. It is important to retain the sign of the values because in the next step the ranks of the positive differences and the negative differences shall be added to obtain (W^+) and (W^-), respectively.

If the sum of the absolute ranks is about 20 or above, a normalization (Gaussian approximation) can be used:

$$\mu_W = \frac{n(n+1)}{4} \tag{133}$$

and

$$\sigma_W = \sqrt{\frac{n \times (n+1) \times (2n+1)}{24}} \qquad (134)$$

The difference can then be expressed as a z-score:

$$z = \frac{W_{\max} - \mu_W}{\sqrt{\frac{n \times (n+1) \times (2n+1)}{24} - q}} \qquad (135)$$

The z-value can be evaluated using an ordinary t-table, but it is often suggested to use $z > 1.96$ irrespective of the number of observations.

The ranking may include ties. In principle, there are two types of ties, one when the difference is zero and the other when the differences are the same. In evaluating the signed-rank test, all differences that are zero should be disregarded. Other ties are given the mean of their ranks, e.g., suppose there are two differences with the rank of 11 then they would each be given the rank of 11.5 and the next rank 13. If there were three, they would occupy the ranks 11, 12, and 13, each being given 12, and the next rank 14. EXCEL does not handle the ranks this way but would assign the rank of 11 to all in both examples. In EXCEL 2010 there are two functions RANK.EQ and RANK. AVG, the first giving each tie the same number the second the average of the numbers of the ties.

It is advisable to check the assignment of ranks. The sum of the possible ranks is always equal to

$$\frac{n \times (n+1)}{2} \qquad (136)$$

i.e., the same as unique numbers in a "cross-table".

The calculation of the z-value thus disregards any differences that are zero and compensates the variance in the denominator by subtracting $q = \frac{t^3 - t}{48}$ for each group of tied values, where t is the number of ties in each group. In laboratory practice, it is rarely necessary to make this compensation.

There are different procedures to evaluate the comparison. The above may be inappropriate for small numbers when instead a table is necessary. Normally, the smaller of the ranked sums is compared with the table value. A rule of thumb is that the larger the difference between the W^+ and W^-, the more probable is the significance of the difference.

Example

The following paired data are extracted from a comparison of two measuring procedures for S-Triglycerides, known for belonging to a skew distribution:

1. Calculate the difference as a decrease between what is appointed as the first measurement and the second but keep track of the sign. Thus, a decrease is a negative number.
2. Ignore the sign and rank the absolute difference. Any differences that are 0 should be completely disregarded.
3. Ties should be given the same numbers and the remaining updated accordingly.
4. Find the sum of the positive ranks and the negative ranks.

	1	2	3	4	5	6	7	8	9	10	11	12	13	14	15
Assay 1	1.24	1.34	1.39	1.41	1.64	1.44	1.48	1.51	1.54	1.54	1.54	1.62	1.63	1.65	1.70
Assay 2	1.30	1.50	1.70	1.50	1.44	1.47	1.60	1.60	1.80	1.50	1.70	1.90	1.81	1.70	1.65
Diff	-0.06	-0.16	-0.31	-0.09	0.20	-0.03	-0.12	-0.09	-0.26	0.04	-0.16	-0.28	-0.18	-0.05	0.05
Abs diff	0.06	0.16	0.31	0.09	0.20	0.03	0.12	0.09	0.26	0.04	0.16	0.28	0.18	0.05	0.05
Rank	5	9.5	15	6.5	12	1	8	6.5	13	2	9.5	14	11	3.5	3.5

Thus, the (W^+) is 17.5 and the (W^-) 102.5; $(W^+)+(W^-)$ is

$$\frac{12 \times (12+1)}{2} = 120$$

$$\mu_W = \frac{15(15+1)}{4} = 60; \quad \sigma_W = \sqrt{\frac{60 \times 31}{6}} = 17.61;$$

$$q = \sum \frac{t^3 - t}{48} = \frac{2^3 - 2 + 2^3 - 2 + 2^3 - 2}{48} = 0.375;$$

$$z = \frac{102.5 - 60}{\sqrt{17.61^2 - 0.375}} = 2.42$$

Thus, the difference between the pairs would be judged significant with a $p < 0.05$ ($p = 0.016$; two-tailed).

Mann-Whitney Test for Unpaired Samples

The nonparametric method for unpaired samples is the Mann-Whitney test or Mann-Whitney U-test and thus the nonparametric solution to evaluating two independent data sets comparable to the Student's t_{ind}. The test can be described as ranking all the results as if they belonged to one measurement and then sum the ranks of the samples belonging to the two methods, separately. The sum of the ranks is T_S and T_L, representing the smaller and larger rank sums, respectively. If the groups were the same, the difference in rank sums should be small. The probability of the sums being equal is evaluated by the binominal theorem.

There are different ways to evaluate the difference:

$$U = T - \frac{n_a \times (n_a + 1)}{2} \tag{137}$$

where n_a refers to the number in the group with the lower rank sum. U is the test statistic if estimated from the group with the lower rank sum. Consult a "Mann-Whitney table" which usually has the number of observations in both groups as entries. The target is to find the last "probability column" that does not contain the statistic (U). There are different layouts of the tables, but they only cover up to a total number of observations of about 20. If there are more observations, the rank sum approaches a normal distribution and the z-value can be calculated. Two approaches are available:

$$\mu_S = \frac{n_S(N + 1)}{2} \quad \text{(Altman)} \tag{138}$$

Alternatively,

$$\mu_S = \frac{n_S n_L}{2} \quad \text{(Engineering handbook)} \tag{139}$$

$$\sigma_S = \sqrt{\frac{n_S \times n_L \times (N + 1)}{12}} \tag{140}$$

$$z = \frac{T_S - \mu_S}{\sigma_S} \quad \text{(Altman)} \tag{141}$$

$$z = \frac{U_S - \mu_S}{\sigma_S} \quad \text{(Engineering handbook)} \tag{142}$$

where n_S is the number of observations in the group with the lower rank, n_L the number of observations in the other group, T_S is the lower rank sum, and U_S is the statistic calculated according to Equation (137). The z-value can then be evaluated by a standard normal distribution or $NORMSDIST(z)$.

Example

The concentration of two different materials was measured by the same procedure:

	1	2	3	4	5	6	7	8	9	10	11	12	
Material 1	1.24	1.34	1.39	1.41	1.64	1.44	1.48	1.51	1.54	1.54	1.54	1.62	
Material 2	1.30	1.50	1.70	1.50	1.44	1.47	1.60	1.60	1.80	1.50	1.70	1.90	Sum
Rank	1	4	5	6	3	7	10	14	15	16	17	19	117
	2	12	21.5	12	8	9	20	18	23	12	21.5	24	183

$$U_S = 12 \times 12 + 0.5 \times 12 \times 13 - 117 = 105;$$

$$U_S + U_L = \frac{24 \times (25 + 1)}{2} = 300$$

$$\mu_S = \frac{12 \times (24 + 1)}{2} = 150 \quad \text{(Altman)}$$

$$\mu_S = \frac{12 \times 12}{2} = 72 \quad \text{(Engineering handbook)}$$

$$\sigma_S = \sqrt{\frac{12 \times 12 \times (24 + 1)}{12}} = 17.32$$

$$z = \frac{|117 - 150|}{17.32} = 1.91 \quad \text{(Altman)}$$

The $NORMSDIST$ gives $p=0.03$:

$$z = \frac{|105 - 72|}{17.32} = 1.90 \quad \text{(Engineering handbook)}; p = 0.03$$

The probability can also be estimated from the test statistic (U). A selection of lines from an appropriate Mann-Whitney table indicates the probability:

n_1	n_2	$p=0.1$	$p=0.05$	$p=0.02$	$p=0.01$
11	12	104-160	99-165	94-170	90-174
12	12	120-180	115-185	109-191	105-195
12	13	125-187	119-193	113-199	109-203

In this example, there are 12 observations in each group and the test statistic, U_S, is 105. The last column of the table that *does not* include the test statistic is $p=0.02$.

REGRESSION

Regression is the statistician's term for describing the relation (dependence, association) between two variables. By convention, the independent (reference or comparative) variable is shown on a horizontal axis (the X-axis) and the dependent (test) variable on the vertical (Y-axis). It may be useful to visualize the independent variable as the cause and the dependent as the effect variable. This is a common terminology particularly in multivariate analysis but that does not imply that the regression will address the causality of a found association.

In analytical work, regression analysis is used for calibration functions. Regression is also used to compare the results of two measurement procedures.

To establish a regression function, the two quantities are measured in the same sample and thus pairs of values will be obtained which can be represented in a two-dimensional diagram, often recognized as a "scattergram" (Figure 6). The simplest regression function describes a linear (first order) relationship, but there are innumerable types of functions. The mathematical function that describes a linear relationship (regression) can be established from a minimum of two pairs of observations. Axiomatically, one and only one straight line can be drawn between two points.

It is recommended in comparisons to display observations in a scattergram to provide a visual impression of the data set. This will facilitate recognizing trends, "outliers" and distribution of data points.

Ordinary Linear Regression

The ordinary linear regression (OLR) or linear least square regression function describes a straight line in a two-dimensional diagram. Its mathematical representation is

$$Y = b \times X + a \tag{143}$$

where b is the slope of the line and a is the Y-intercept, i.e., the Y-value where the line crosses the Y-axis, i.e., when the value of Y if $X=0$.

The OLR establishes a line which minimizes the vertical differences between each observation and the line and disregards the relation to the x value. Therefore, it is important to choose the variable with the smaller measurement uncertainty as the independent quantity on the horizontal axis (x-value). The regression line is a kind of average of all observations in the measuring interval. It is always centered on the average of the independent and average of dependent variables $(\bar{x}/\bar{y})$.

If a set of paired observations (x_i/y_i) and the number of pairs (n) are given, the regression function can be calculated.

First the slope $(b_{y/x})$ or "regression coefficient" is calculated:

$$b_{y/x} = \frac{\sum_{i=1}^{n}[(x_i - \bar{x}) \times (y_i - \bar{y})]}{\sum_{i=1}^{n}(x_i - \bar{x})^2} = \frac{\sum_{i=1}^{n}(x_i \times y_i) - n \times \bar{x} \times \bar{y}}{\sum_{i=1}^{n}x_i^2 - n \times (\bar{x})^2} \tag{144}$$

The formulas can be simplified by defining the "sum of squares":

$$SS_{xx} = \sum_{i=1}^{n}x_i^2 - \frac{\left(\sum_{i=1}^{n}x_i\right)^2}{n} = \sum_{i=1}^{n}(x_i - \bar{x})^2 =$$
$$(n_x - 1) \times s(x)^2 \tag{145}$$

$$SS_{yy} = \sum_{i=1}^{n}y_i^2 - \frac{\left(\sum_{i=1}^{n}y_i\right)^2}{n} = \sum_{i=1}^{n}(y_i - \bar{y})^2 =$$
$$(n_y - 1) \times s(y)^2 \tag{146}$$

$$SS_{xy} = \sum_{i=1}^{n}(x_i \times y_i) - \frac{\sum_{i=1}^{n}x_i \times \sum_{i=1}^{n}y_i}{n} =$$

$$\sum_{i=1}^{n}(x_i - \bar{x}) \times (y_i - \bar{y}) \tag{147}$$

Then,

$$b_{y,x} = \frac{SS_{xy}}{SS_{xx}} \tag{148}$$

The means of the variables are assumed to be on the line and thus satisfy the function. The pair of means is also recognized as the centroid of the function. The means of the results of the test $(\bar{y})$ and reference $(\bar{x})$ measurements can thus be used, together with the estimated slope, to estimate the intercept and thus the regression function can be calculated:

$$\bar{y} = b_{y/x} \times \bar{x} + a; \quad a = \bar{y} - b_{y/x} \times \bar{x} \tag{149}$$

Since the means of the quantity values are used in the calculation of the slope and also the intercept, it is important that the means are representative of the quantity values. This, strictly, requires that the quantity values are normally distributed along both axes.

In the OLR, the sum of the squared vertical distances between the observations (y) and the point calculated by the regression function on the regression line $(\hat{y})$ are minimized. These distances are recognized as "residuals."

A consequence of the OLR model is that it does not include, or consider, any measurement uncertainty of the independent variable.

Many quantitative relations between a signal and a concentration in analytical chemistry are linear, and the OLR is frequently used in calibrations. There are many exceptions from linearity in measuring systems, e.g., the relation between signal and concentration in immunoassays is rarely linear.

A calibration is usually performed and displayed with the signal on the Y-axis and the concentration on the X-axis. The concentration of the calibrator is usually known with a small, negligible, or zero uncertainty, and therefore, the OLR is a suitable model for calibration if the regression is linear. When the calibration is used to convert a signal to concentration, its reverse is used, i.e., the signal (Y) is entered to the calibration function $X = \dfrac{Y - a}{b_{y/x}}$.

The estimated concentration will thus be traceable to the calibrator via the calibration function.

The uncertainty of the signal is transferred to the estimated concentration, and the result will have an increased uncertainty compared to that of the calibrator value.

The uncertainty (standard error) of the slope and intercept can be estimated by formulas of several different formats, giving identical results:

$$u\left(b_{y/x}\right) = \sqrt{\frac{\sum_{i=1}^{n}\left(y_i - \hat{y}_i\right)^2}{\sum_{i=1}^{n}(x_i - \bar{x})^2} \times \frac{1}{(n - 2)}} = \frac{s_{y,x}}{\sqrt{\sum_{i=1}^{n}(x_i - \bar{x})^2}} =$$

$$\frac{s_{y,x}}{\sqrt{(n - 1) \times s(x)^2}} = \frac{s_{y,x}}{\sqrt{SS_{xx}}} \tag{150}$$

where $\hat{y}_i$ is the value of the dependent variable estimated from the corresponding x_i and the regression function and thus $\left(y_i - \hat{y}_i\right)$ is the residual.

$s_{y,x}$ is the standard deviation of the residuals

$$s_{y,x} = \sqrt{\frac{\sum_{i=1}^{n}\left(y_i - \hat{y}_i\right)^2}{n - 2}} \text{ (see Equation 164).}$$

Note The uncertainty that is obtained corresponds to the standard error of the slope. Therefore, the CI of the slope is $\pm z$ x $u(b)$.

The significance of the slope being different from zero (i.e., horizontal)—or indefinite [1] (i.e., vertical)—is obtained by calculating the Student's independent t-value (126). The standard error of these extremes is zero and the calculation of the t-value simplified as

$$t = \frac{b_{y/x} - 0}{\sqrt{u(b)^2 - 0}} = \frac{b_{y/x}}{u(b)} \text{ and } \frac{b_{y/x} - 1}{u(b)}, \text{ respectively} \tag{151}$$

The t-value is evaluated using an ordinary t-table and $df = n - 2$. At low values of the slope ($b_{y/x}$), it may be important to demonstrate that the slope is different from zero; if not, all x-values give the same y-value, i.e., $Y = a$.

Note The correlation of the variables can be significant, and the coefficient of variation high even if the slope is not significantly different from zero. If the slope is zero, however, there cannot be an association between the variables or quantities.

The uncertainty (standard error) of the intercept includes the uncertainty of the residuals, $s_{y,x}$ (164). The formula comes in many forms:

$$u(a) = s_{y,x} \times \sqrt{\frac{\sum_{i=1}^{n} x_i^2}{n \times (n-1) \times s(x)^2}} = s_{y,x} \times \sqrt{\frac{\sum_{i=1}^{n} x_i^2}{n \times \sum_{i=1}^{n} (x_i - \bar{x})^2}} =$$

$$s_{y,x} \times \sqrt{\frac{1}{n} \times \frac{\sum_{i=1}^{n} (x_i)^2}{SS_{xx}}} \tag{152}$$

If the data are displayed in an EXCEL spreadsheet, the OLR can be directly shown in the graph, by adding a "trendline." There are also functions to directly calculate the slope and intercept for a data set, e.g., *SLOPE(Y1:Yn,X1:Xn)* and *INTERCEPT(Y1:Yn,X1:Xn)*, where (*Y1:Yn*, *X1:Xn*) defines the data set.

Note A characteristic of the OLR, or "method of linear least squares," is that the variance (uncertainty) of measurements of the independent quantity (X value) is assumed to be zero or that the ratio between $s(y)^2$ and $s(x)^2$, referred to as λ_i, discussed below (153), is large. The OLR is relatively robust, and acceptable results may be obtained also if the variance of X values is >0. Further, the variance of the independent variable (Y) should be homoscedastic, i.e., the same within the measuring interval.

The OLR is sensitive to outliers, and extreme values will therefore have a major impact on the OLR function (see section on "Leverage").

The quantity values should ideally be normally distributed in both directions and the OLR is regarded as rather robust also in that respect.

Example

Estimate the OLR of the following data

	1	2	3	4	5	6	7	8	9	10	11	12	Mean	s
X-value	1.24	1.34	1.39	1.41	1.64	1.44	1.48	1.51	1.54	1.54	1.54	1.62	1.47	0.12
Y-value	1.30	1.50	1.70	1.50	1.44	1.47	1.60	1.60	1.80	1.50	1.70	1.90	1.58	0.17

From EXCEL, the slope and intercept were 0.7856 and 0.4261, respectively. The regression function can also be displayed on the graph as a "trendline" function (Figure 7).

Using the formulas presented:

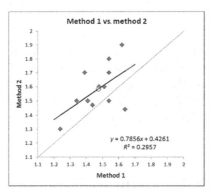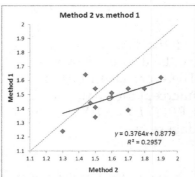

FIGURE 7   Scatterplot of results of the example. In the right panel, the variables have been swapped, i.e., the x-values of the table are on the vertical axis and the y-values are on the horizontal axis. The function of the OLR is shown and the coefficient of detection (r^2). *Note*: The functions are different but describe the same relation, i.e., the y dependence on x in both examples. The coefficient of determination is unchanged. The equal-sized units and length of axes facilitate comparing the regressions with the "equal line" with a slope of 1, i.e., 45°. The regression lines pass through the average of the values of the dependent and independent variables.

$SS_{xx} = \sum_{i=1}^{n}(x_i - \bar{x})^2 = (n-1) \times s(x)^2 = 11 \times 0.12^2 = 0.148;$

$SS_{yy} = 11 \times 0.17^2 = 0.309$

$SS_{xy} = \sum_{i=1}^{n}(x_i - \bar{x}) \times (y_i - \bar{y}) = 0.116$

$b_{y/x} = \dfrac{SS_{xy}}{SS_{xx}} = 0.785; \quad b_{x/y} = \dfrac{SS_{xy}}{SS_{yy}} = 0.386$

where $b_{y/x}$ is the slope of the regression function "Y on X." The slope of "X on Y" is written $b_{x/y}$. If the slope is not identified by an index, it is usually $b_{y/x}$:

$a = \bar{y} - b \times \bar{x} = 1.58 - 0.785 \times 1.47 = 0.426$

The uncertainty of the slope:

$u(b) = \dfrac{s_{y,x}}{\sqrt{SS_{xx}}} = \dfrac{0.148}{\sqrt{0.148}} = 0.383$ ($s_{y,x}$ is 0.148; see Equation 164)

The uncertainty of the intercept is

$u(a) = s_{y,x} \times \sqrt{\dfrac{1}{n} \times \dfrac{\sum_{i=1}^{n}(x_i)^2}{SS_{xx}}} = 0.148 \times \sqrt{\dfrac{26.226}{12 \times 0.148}} = 0.567$

Deming Regression

In practical work, there is usually a variation in the measurements also of the independent variable. The Deming linear regression minimizes the perpendicular (*ortho*) distances between the observations and a calculated regression line (Figure 8) by including the ratio between the variance of the independent and dependent observations. If the ratio is equal to 1, then the model minimizes the perpendicular distance to the regression line. This is the orthogonal regression. The larger the ratio, the more vertical the minimal distance will be and at high ratios eventually becomes vertical and the regression function becomes identical to the OLR.

Before the slope (b_D) can be calculated, the ratio between the measurement variance of the quantities $s(y)^2$ and $s(x)^2$ must be defined

$\lambda_i = \dfrac{[s(y)]^2}{[s(x)]^2}$ (153)

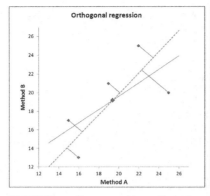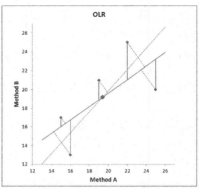

FIGURE 8 Deming (dotted) and ordinary linear regression (solid) lines. Thin solid lines between the observations and the regression lines represent the distances that are minimized. In the left panel the lambda (λ_i) value is 1 resulting in minimizing the perpendicular lines (green) to the anticipated regression line (Deming). In the right panel, the (λ_i) $\gg 1$ and accordingly the vertical lines to an assumed regression line are minimized (blue). This is the ordinary linear regression line.

It is an advantage to also calculate a function (V) that occurs repeatedly in the calculations

$$V = \frac{\sum_{i=1}^{i}(y_i - \bar{y})^2 - \lambda_i \times \sum_{i=1}^{i}(x_i - \bar{x})^2}{2 \times \sum_{i=1}^{i}[(x_i - \bar{x}) \times (y_i - \bar{y})]} = \frac{SS_{yy} - \lambda_i \times SS_{xx}}{2 \times SS_{xy}}$$

(154)

$$b_D = V + \sqrt{V^2 + \lambda_i}$$

(155)

Note In statistical literature, the λ is often defined as $\dfrac{[s(x)]^2}{[s(y)]^2}$, i.e., $\lambda_i = 1/\lambda$. If that definition is used, then the corresponding changes in Equations (154) to (158) shall be made.

The Deming regression approaches the OLR if the $s(y)$ is much larger than $s(x)$ and accordingly $\lambda_i \gg 1$. It may therefore be more convenient to use λ_i as defined in Equation (153) in these discussions. The Deming regression method, like the OLR, requires that the variance is constant, i.e., homoscedastic, for both variables and the observations reasonably normally distributed.

The b_D can also be calculated from the OLR regression coefficient Y on X and the Pearson correlation coefficient (r). This is achieved by introducing $b_{y/x}$ and $b_{x/y}$ and rearranging Equation (154)

$$V = \frac{SS_{yy} - \lambda_i \times SS_{xx}}{2 \times SS_{xy}} = \frac{SS_{yy}}{2 \times SS_{xy}} - \frac{\lambda_i \times SS_{xx}}{2 \times SS_{xy}} =$$

$$\frac{\dfrac{1}{2 \times SS_{xy}}}{SS_{yy}} - \frac{\dfrac{\lambda_i}{2 \times SS_{xy}}}{SS_{xx}} = \frac{1}{2b_{x/y}} - \frac{\lambda_i}{2b_{y/x}} \tag{156}$$

Since

$$\frac{1}{b_{x/y}} = \frac{b_{y/x}}{r^2}; \quad b_{x/y} = \frac{r^2}{b_{y/x}}; \quad r = \sqrt{b_{y/x} \times b_{x/y}} \tag{157}$$

$$V = \frac{b_{y/x}}{2 \times r^2} - \frac{\lambda_i}{2 \times b_{y/x}} \tag{158}$$

which is then entered into Equation (155). The intercept is estimated as described in Equation (149).

Example

Use the same data set as in the previous section, copied here

	1	2	3	4	5	6	7	8	9	10	11	12	Mean	s
X-value	1.24	1.34	1.39	1.41	1.64	1.44	1.48	1.51	1.54	1.54	1.54	1.62	1.47	0.12
Y-value	1.30	1.50	1.70	1.50	1.44	1.47	1.60	1.60	1.80	1.50	1.70	1.90	1.58	0.17

Assume that $s(x) = s(y)$ in this experiment. Thus, $\lambda_i = 1$.

$$\lambda_i = \frac{[s(y)]^2}{[s(x)]^2} = 1$$

$$V = \frac{SS_{yy} - \lambda_i \times SS_{xx}}{2 \times SS_{xy}} = \frac{0.309 - 1 \times 0.148}{2 \times 0.116} = 0.692$$

$$b_D = V + \sqrt{V^2 + \lambda_i} = 0.692 + \sqrt{0.692^2 + 1} = 1.90$$

$$a_D = \bar{y} - b_D \times \bar{x} = 1.58 - 1.58 \times 1.47 = -1.29$$

The alternative calculation of the slope is

$$\frac{1}{b_{x/y}} = \frac{b_{y/x}}{r^2}; \quad b_{x/y} = \frac{r^2}{b_{y/x}} = \frac{0.296}{0.786} = 0.377$$

$$V = \frac{b_{y/x}}{2 \times r^2} - \frac{\lambda_i}{2 \times b_{y/x}} = \frac{0.786}{2 \times 0.296} - \frac{1}{2 \times 0.786} = 0.692$$

etc.

When performing these calculations, it is essential to retain as many value digits as possible and only make any rounding in the final result.

There are different formulas to estimate the uncertainty of the slope u_{bD} and intercept u_{aD} and they do not always give the same result.

The uncertainty of the slope of the Deming regression is

$$u_{bD} = \sqrt{\frac{b_D^2 \times (1 - r^2)}{r^2 \times (n - 2)}} = \frac{b_D}{r} \times \sqrt{\frac{1 - r^2}{n - 2}} \qquad (159)$$

where r is the Pearson correlation coefficient (see Equation 165).

The uncertainty of the intercept:

$$u_{aD} = \sqrt{\frac{(u_{bD})^2 \times \sum_{i=1}^{n} x_i^2}{n}} \qquad (160)$$

In the formulas used here, the r (correlation coefficient) and the sum of the squared results of the independent variable are necessary. They are available in EXCEL: r: CORREL $(Y1{:}Y12{;}X1{:}X12)=0.543$ and $SUMSQ(X1{:}X12)=26.226$, respectively:

$$u_{bD} = \sqrt{\frac{b_D^2 \times (1 - r^2)}{r^2 \times (n - 2)}} = \sqrt{\frac{1.90^2 \times (1 - 0.544^2)}{0.544^2 \times (12 - 2)}} = 1.38$$

$$u_{aD} = \sqrt{\frac{(u_{bD})^2 \times \sum_{i=1}^{n} x_i^2}{n}} = \sqrt{\frac{0.5241 \times 26.23}{12}} = 0.775$$

Weighted Regression

In a regression analysis, it may not be reasonable to assume that every observation should be treated with equal weight. A procedure that treats all of the data equally would give less precisely measured points, e.g., at the ends of the measuring interval, more influence than they should have and would give highly precise points too little influence. To minimize the influence of outliers and extreme values methods for weighting the data have been developed. The reader is referred to textbooks on statistics for further discussions of weighting.

Other Regression Functions

Two-Point Calibration

Between two points, one, and only one, straight line can be drawn. This is utilized in establishing the calibration function from two concentrations x_1/y_1 and x_2/y_2:

$$b_{y/x} = \frac{(y_2 - y_1)}{(x_1 - x_2)} \tag{161}$$

where x_1, x_2 and y_1, y_2 are the corresponding results of the independent and dependent variables, respectively.

The regression (calibration) function is

$$Y - y_1 = b \times (X - x_1) \quad \text{or} \quad Y - y_2 = b \times (X - x_2) \tag{162}$$

The average of the two points (x_1/y_1 and x_2/y_2) may also be used to establish the regression function, instead of either of the measured points.

A "two-point calibration" assumes a linear relation between the quantities.

The "two-point" formula can also be used to establish a recalibration function from patient or control samples.

The Pearson correlation coefficient (r) for a function derived from only two points will, by definition, be 1.

Regression from two intervals

The standard deviations $s(x)$ and $s(y)$ represent the distribution of the values. It is reasonable that the ratio between $s(y)$ and $s(x)$ is a measure of the slope ($b_{y/x}$):

$$b_{y/x} = \frac{s(y)}{s(x)} \tag{163}$$

The intercept is then estimated as above (149) using the coordinates of the average.

Example

In the previous example, the $s(y)$ and $s(x)$ were 0.168 and 0.116, respectively. The estimated $b_{y/x} = 1.45$, intercept $= -0.54$. The usefulness and accuracy of this approximation of $b_{y/x}$ are depending on the underlying distributions which must be normal to ensure that the calculations $s(y)$ and $s(x)$ represent the spread of the data. Great care should be exercised in the use of the method. Since the $s(x)$ and $s(y)$ are the positive roots of the corresponding variances, this method will always result in a $b_{y/x} > 0$.

Compare OLR $Y = 0.786X + 0.426$; Deming $Y = 1.90X - 1.23$.

Bartlett Regression

Order the data set according to the independent variable and divide it into three intervals low, mid and high with equal numbers of observations. If the number of observations is not a function of three, adjust the mid interval so the low and high include the same number of observations. Calculate the average of the high and low interval and calculate the slope according to the two-point formula (161) and the intercept from the average of all the y and x-values, respectively. The Bartlett's regression is assumed to allow a measurement uncertainty in both dependent and independent variables.

Passing-Bablok Regression

Other techniques have been developed to accommodate variances in the results of both variables, and Passing-Bablok regression, similar to the Thiel-Sen estimator, is the most favoured. These are nonparametric and do thus not assume any particular distribution. Essentially, the Passing-Bablok (P-B) calculates the slope for all possible lines combining the observations, excluding those which are 0 or indefinite. The slope of the regression line for all observations (b) is the median of those of all the connecting lines. The intercept (a) is then calculated from the median or mean of all observations.

As in calculations based on medians, the influence of outliers is less than for parametric methods.

The Passing-Bablok regression requires comparatively extensive calculations.

Example

Use the data in the previous example; calculate the average of the low and high thirds. The slope (*b*) and intercept (*a*) of the other described functions are summarized.

Note The different regression functions are designed to handle measurement uncertainties differently which is not reflected in the used dataset.

	Low	High	All		Bartlett	OLR	DLR	P-B
Avg (x)	1.345	1.500	1.474	b	0.56	0.79	1.90	1.67
Avg (y)	1.585	1.635	1.584	a	0.76	0.43	−1,29	−0.86

The regression lines are displayed in Figure 9.

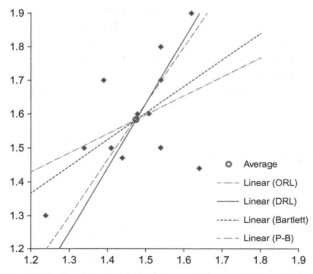

FIGURE 9   Regression lines calculated according to ordinary linear regression (OLR), Deming regression (DLR), Bartlett and Passing-Bablok, from the same dataset.

Linearity

Linearity is characterized by a first-order function that describes the relation between a signal, e.g., light absorbance and the concentration of an analyte.

If the intercept is zero, the function will provide results that are directly proportional to the concentration of the analyte in the sample. A sample that is diluted $1+1$ shall therefore give a signal that is half of the original signal and the result shall be half of that the original sample. This is not always the case in biological samples; a reason may be that inhibitors are inefficient or more potent in diluted samples.

If two methods pertaining to measure the same quantity are linearly related, they may be assumed to in fact measure the same quantity irrespective of the numerical results. This makes recalibration using a reference procedure possible.

A thorough procedure for evaluate the linearity of a measurement procedure has been published as EP6 by the CLSI (www.CLSI.org).

Higher Order Regressions

It is not uncommon that a relation between quantities is nonlinear, i.e., the function that describes the relation is of a higher order. Because linear regressions are easy to apply, analysts usually try to linearize the functions, as has been described above (Table 2 and Figure 2).

Higher order functions may be fitted to data by special procedures which are available in some statistical packages. EXCEL offers five different trendlines to be fitted to tabled data. They are available by activating a data set in a graph "FORMATE DATA SERIES."

Residuals

Intuitively, the spread is related to the distances between the observations and the estimated regression line. These distances are the "residuals" and their standard deviation $(s_{y,x})$ is

$$s_{y,x} = \sqrt{\frac{\sum_{i=1}^{n}(y_i - \hat{y}_i)^2}{n-2}} = \sqrt{\frac{1}{n-2} \times (SS_{yy} - b \times SS_{xy})} =$$

$$\sqrt{\frac{\sum_{i=1}^{n}y_i^2 - n \times y_i^2 - b_{y/x}^2 \times \left(\sum_{i=1}^{n}x_i^2 - n \times (\bar{x})^2\right)}{n-2}} =$$

(164)

$$\sqrt{\frac{(n-1) \times \left(s_y^2 - b_{y/x}^2 \times s_x^2\right)}{n-2}} =$$

$$s_y \times \sqrt{\frac{n-1}{n-2} \times (1 - r^2)}$$

where $b_{y/x}$ is the slope, s_y and s_x the standard deviation of the y and x variables, respectively. $\hat{y}$ (y-roof) is the y-value calculated using the regression function at a particular x_i. The $y_i - \hat{y}$ is also recognized as the "error term."

In EXCEL, the $s_{y,x}$ is calculated by the function $STEYX(y{:}s, x{:}s)$.

The $s_{y,x}$ is also recognized as residual standard deviation (rsd) or (s_{res}), residual standard error (rse), standard deviation of the line (sdl), standard error of the estimate (see), or linear residual standard deviation ($ressd$).

CORRELATION AND COVARIANCE

Correlation describes and quantifies the strength and direction of a linear relationship between two random variables.

Correlation Coefficient

A correlation coefficient describes the dependence of two random variables and thus describes the spread of the observed pairs of observations. This dependence is quantitatively described by the *correlation coefficient*, the Pearson Product-Moment correlation coefficient, which can be calculated by many seemingly different formulas which, however, can be derived from each other but represent different ways to visualize the correlation coefficient:

$$r = \frac{\sum_{i=1}^{i=n}[(x_i - \bar{x}) \times (y_i - \bar{y})]}{\sqrt{\sum_{i=1}^{i=n}(x_i - \bar{x})^2 \times \sum_{i=1}^{i=n}(y_i - \bar{y})^2}} = \frac{SS_{xy}}{\sqrt{SS_{xx} \times SS_{yy}}} =$$

$$\sqrt{\frac{SS_{xy}}{SS_{xx}}} \times \sqrt{\frac{SS_{xx}}{SS_{yy}}} = b_{y/x} \times \sqrt{\frac{SS_{xx}}{SS_{yy}}} =$$

$$\left[b_{y/x} \times \sqrt{\frac{\frac{SS_{xx}}{SS_{xy}}}{\frac{SS_{yy}}{SS_{xy}}}} = b_{y/x} \times \sqrt{\frac{b_{x/y}}{b_{y/x}}} = \sqrt{b_{y/x} \times b_{x/y}} = \right] \tag{165}$$

$$b_{y/x} \times \sqrt{\frac{\sum_{i=1}^{n}(x_i - \bar{x})^2}{\sum_{i=1}^{n}(y_i - \bar{y})^2}} = b_{y/x} \times \sqrt{\frac{s(x)^2}{s(y)^2}} = b_{y/x} \times \frac{s(x)}{s(y)}$$

SS_{xy}, SS_{xx}, and SS_{yy} are defined in Equations (145)–(147).

The formula can also be written in yet another form that avoids calculation of the means:

$$r = \frac{\sum_{i=1}^{n}x_i \times y_i - \frac{\sum_{i=1}^{n}x_i \times \sum_{i=1}^{n}y_i}{n}}{\sqrt{\left(\sum_{i=1}^{n}x_i^2 - \frac{\left(\sum_{i=1}^{n}x_i\right)^2}{n}\right) \times \left(\sum_{i=1}^{n}y_i^2 - \frac{\left(\sum_{i=1}^{n}y_i\right)^2}{n}\right)}} \tag{166}$$

r can assume values between (-1) and $(+1)$, i.e., $-1 \leq r \leq +1$ or $r \leq |1|$.

All calculations of r include the mean, the standard deviation, or derivatives thereof and thus require that the data are normally or close to normally distributed. The correlation coefficient describes the scatter of the observations or the association between the variables. Thus an $r=0$ indicates no association and $r=1$ a perfect direct association, whereas $r=-1$ indicates a perfect inverse association (Figure 10).

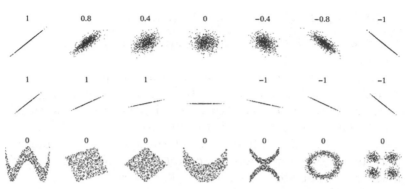

FIGURE 10   The Pearson correlation coefficient calculated for differently distributed data sets. The top row illustrates that the correlation coefficient describes the scatter, and its sign the slope of a linear regression (middle). The many non parametric data sets in the bottom row have a correlation coefficient of 0. *From Wikipedia Commons.*

The correlation coefficient can be calculated for any data set and only describes the spread of the observations in a two-dimensional scatter plot. The relation of r to the OLR function is formal as shown in Equation (165). This does not exclude a relation between the correlation coefficient and the linear regression, and it will be sensitive only to a linear relationship between two variables (which may exist even if one is a non-linear function of the other).

If the variables are expressed as vectors, the $\cos(\varphi)$, where φ is the angle between vectors, will be equal to r. $\cos(\varphi)$ is equal to

$$\cos(\varphi) = \frac{\sum_{i=1}^{i=n}[(x_i - \bar{x}) \times (y_i - \bar{y})]}{\sqrt{\sum_{i=1}^{i=n}(x_i - \bar{x})^2 \times \sum_{i=1}^{i=n}(y_i - \bar{y})^2}}$$

which is already identified as one of the definitions of r (165).

If the slope is calculated according to Equation (161), then $r=1$ which would be expected since the distribution of the data has been simplified to one number, representing an interval. It is an axiom that between two points, only one straight line can be drawn. Consequently, the r will be $=\pm1$.

Even a high correlation coefficient does not imply a causal relationship between the variables.

Coefficient of Determination (Pearson)

The square of the correlation coefficient (r^2), also known as the *coefficient of determination*, is understood as the fraction of the variation in y_i that is accounted for by a linear fit of x_i to y_i; or differently expressed is the proportion of variance in common between the two variables. For example, for $r=0.7$, $r^2=0.49$ and thus, only 49 % of the variation is explained by the linear fit. The coefficient of determination is the quantity that should be interpreted; the correlation coefficient overestimates the association between the variables.

The relation between the residual standard deviation and the correlation coefficient and coefficient of determination can be approximated:

$$s_{y/x} = s(y) \times \sqrt{\frac{n-1}{n-2} \times (1-r^2)}$$

$$\left(\frac{s_{y,x}}{s(y)}\right)^2 = \frac{n-1}{n-2} \times (1-r^2)$$

(167)

If $n-1$ approaches $n-2$, then

$$r^2 = 1 - \left(\frac{s_{y,x}}{s(y)}\right)^2 \qquad (168)$$

Consequently, the smaller the ratio between $s_{y,x}$ and $s(y)$ the larger r^2. This relation is important to observe in evaluation of a comparison of results by regression analysis.

It should be stated that correlation does not equal causation as already pointed out. There are many reasons to be careful drawing conclusions from correlation coefficients; even if the p-value indicates a high degree of probability—or significance—this may not give a clue to the root cause. It is essential also to view the correlation in relation to the regression, e.g., in a scatterplot. As will be shown below, the significance of r is highly depending on the number of observations.

Example

The data set used as an example in the section on "Regression" is used previously.

	1	2	3	4	5	6	7	8	9	10	11	12	Mean	s
X-value	1.24	1.34	1.39	1.41	1.64	1.44	1.48	1.51	1.54	1.54	1.54	1.62	1.47	0.12
Y-value	1.30	1.50	1.70	1.50	1.44	1.47	1.60	1.60	1.80	1.50	1.70	1.90	1.58	0.17

The correlation coefficient, r, according to EXCEL is 0.544. Let us apply the first and last expressions in the chain of the algebraically different formulas above (165) which give identical results:

$$\frac{\sum_{i=1}^{i=n}[(x_i - \bar{x}) \times (y_i - \bar{y})]}{\sqrt{\sum_{i=1}^{i=n}(x_i - \bar{x})^2 \times \sum_{i=1}^{i=n}(y_i - \bar{y})^2}} = \frac{0.116492}{\sqrt{0.148292 \times 0.309492}} =$$

$$\frac{0.1165}{0.2142} = 0.544$$

$$b_{y/x} \times \frac{s(x)}{s(y)} = 0.7855 \times \frac{0.1161}{0.1677} = 0.544$$

Spearman Rank Correlation

The *Pearson product-moment correlation*, r, assumes Gaussian distributed data. If this is not the case, the *Spearman's rank correlation* is used to test the direction and strength of the relationship between two variables.

Spearman's rank correlation (r_s) or ρ (rho).

The Spearman rank correlation coefficient is a nonparametric correlation coefficient. It only addresses the ranks of independently ranked variables. Calculating the Pearson correlation from data ranked in ascending order will give an approximate value of the r_s.

The r_s can also be estimated according to

$$\rho \approx r_s = 1 - \frac{6 \times \sum_{i=1}^{n}d_i^2}{n \times (n^2 - 1)} \tag{169}$$

where n is the number of pairs and d_i the difference between the ranks.

If ties occur in either of the data sets, Equation (169) should, in theory, not be used unless the ties are resolved. However, the effect of a few ties is usually small.

Resolving ties means that they are given different ranks. In EXCEL is

$$IF(ISNUMBER(CR), ((RANK(CR, C\$R_1:C\$R_n, 0)$$

$$+COUNT(C\$R_1:C\$R_n) - RANK(CR, C\$R_1:C\$R_n, 1)$$

$$+1)/2), "\cdot") \tag{170}$$

where C denotes Column, R Row, R_1 the first observation, and R_n the last observation. In EXCEL 2010 the function RANK.AVG(CR,C\$R1:C\$Rn,1) will give the same effect.

Example

Assume a set of pair-wise observations:

Obs. 1	Obs. 2	Rank 1	Rank 2	Diff., d_i
88	105	4	8	−4
94	93	9	3	6
83	69	2	1	1
91	91	7	2	5
90	107	6	9	−3
89	100	5	7	−2
82	96	1	5	−4
93	99	8	6	2
83	95	2	4	−2
102	110	10	10	0

Calculate the ranks in increasing order, for instance using the EXCEL function: $RANK(R_i, R_1:R_n, 0)$.

Applying the Pearson product moment correlation to the ranks estimates the r_s to 0.33.

Alternatively, calculate the difference between the ranks (d_i) and apply to Equation (169):

$$r_s = 1 - \frac{6 \times \sum_{i=1}^{n} d_i^2}{n \times (n^2 - 1)} = 1 - \frac{6 \times 115}{10 \times 99} = 1 - 0.697 = 0.30$$

Standard statistical programs, e.g., SigmaPlot, Prism/Graphpad, Statistica, JMP return $r_s = 0.30$.

Significance of r

The standard error of the correlation coefficient is

$$se(r) = \frac{1 - r^2}{\sqrt{n - 2}} \tag{171}$$

where n is the number of individuals (samples, i.e., pairs) in the data set. Strictly, this should only be used for large samples (>100).

The significance of a correlation between two variables is estimated by Student's t-test:

$$|t| = r \times \sqrt{\frac{n - 2}{1 - r^2}}; \quad df = n - 2 \tag{172}$$

The t-value is evaluated by the usual t-distribution table and thus allows the estimation of the statistical significance of r.

Note High values of t, signaling statistical significance, will be obtained with a large number of observations even if r and thus r^2 are small, indicating a limited explanation by the correlation. The interpretation of the r will vary depending on the context. Thus, $r = 0.8$ may be regarded as unsatisfactorily low when comparing two measurement procedures in chemistry or physics, whereas it might be very differently appreciated in social or medical correlation studies where confounding factors may be more abundant.

The CI for (r) is estimated after Fisher's transformation of the r:

$$Z = \frac{1}{2} \times [\ln(1 + r) - \ln(1 - r)] = \frac{1}{2} \times \ln\left(\frac{1 + r}{1 - r}\right) \tag{173}$$

EXCEL offers a function for direct calculation of Z: *FISHER(r)*.

The standard error of Z is

$$s_Z = \pm \frac{1}{\sqrt{n-3}} \tag{174}$$

Thus, the CI will be

$$CI_Z = Z \pm z \times \frac{1}{\sqrt{n-3}} = \pm z \times s_Z \tag{175}$$

where z corresponds to the confidence level, e.g., 1.96 for 95 % confidence level.

Note The difference between capital Z and low case z!

The endpoints of the CI_Z ($-(Z \times s_Z)$ and $+(Z \times s_Z)$, respectively) are entered into Equation (173) to define the confidence limits of r.

Note The CI_r is not symmetrical around r.

$$r_{\lim} = \frac{e^{2z} - 1}{e^{2z} + 1} \tag{176}$$

Example

The correlation coefficient, r, for a linear regression of 35 values was 0.91. Calculate the 95 % CI for r.

Fisher's Z-value (173) is 1.53. $s_Z = \pm 1.96 \times 0.18 = \pm 0.35$ (171). Thus, the CI_Z: 1.18–1.88 (175), corresponding to CI_r: 0.83–0.95 when the endpoints of CI_Z are inserted into Equation (176).

Note The CI is asymmetric around the estimated r.

Covariance

The dependence between two random variables is also described by the covariance which is a measure how two variables vary together.

The covariance can be derived from the calculation of variance (23)

$$s^2 = \frac{\sum_{i=1}^{i=n}(x_i - \bar{x})^2}{n-1}$$ which can be written as

$$s^2 = \frac{\sum_{i=1}^{i=n}(x_i - \bar{x}) \times (x_i - \bar{x})}{n-1}.$$

The covariance (sample) between x and y is then

$$\text{covariance} = \frac{\sum_{i=1}^{i=n}[(x_i - \bar{x}) \times (y_i - \bar{y})]}{n-1} = \frac{1}{n-1} \times SS_{xy} =$$

$$\frac{r \times \sqrt{SS_{xx} \times SS_{yy}}}{n-1} \tag{177}$$

Thus

$$r = \frac{(n-1) \times covariance}{\sqrt{SS_{xx} \times SS_{yy}}} = \frac{(n-1) \times covariance}{\sqrt{(n-1) \times var(x_i) \times (n-1) \times var(y_i)}} \tag{178}$$

and

$$r = \frac{covariance}{\sqrt{var(x_i) \times var(y_i)}} = \frac{covariance}{s(x_i) \times s(y_i)} \tag{179}$$

It is always safe to use the sample variance to estimate the correlation coefficient. The correlation coefficient is therefore a quantity derived from the covariance and is also expressed as the normalized covariance.

The importance of covariance can be shown by an example: Suppose we have two data sets A and B and know the variances of the data sets and their covariance.

Then the variance of $A+B$;

$$VAR(A+B) = VAR(A) + VAR(B) + 2 \times COVAR(A,B).$$

The covariance has a great influence. Thus, when the number of samples reaches 200, the difference between using the population and sample covariance is about 1 %.

It should be noted that the covariance in Equations (177)–(179) refers to the covariance between observations (x- and y-values) not to the covariance between the slope and intercept of the regression function.

The covariance between the slope (a) and the intercept (b) is

$$\text{cov}(a,b) = -\bar{X} \times u\left(b_{y/x}\right)^2 \tag{180}$$

Correlation and Covariance Matrix

Correlation and covariance can only be calculated between two variables at a time. However, if there is a number of data sets, it may be convenient to calculate the correlation and covariance between all of them. The outcome is a correlation/covariance matrix. The EXCEL has innate functions (Figure 11) that can be found under the Add-in—Data analysis, CORRELATION, and COVARIANCE, respectively. These commands directly calculate the desired matrix comprising the covariance or correlation between all the columns or rows, as specified.

Note In the covariance matrix, the diagonal displays the variances of the data in the columns (174) and the other the covariance terms.

The correlation coefficient is independent of the number of observations (165), whereas the covariance includes the

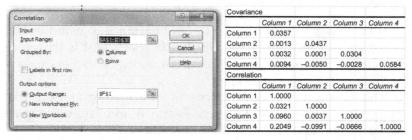

Covariance				
	Column 1	Column 2	Column 3	Column 4
Column 1	0.0357			
Column 2	0.0013	0.0437		
Column 3	0.0032	0.0001	0.0304	
Column 4	0.0094	−0.0050	−0.0028	0.0584
Correlation				
	Column 1	Column 2	Column 3	Column 4
Column 1	1.0000			
Column 2	0.0321	1.0000		
Column 3	0.0960	0.0037	1.0000	
Column 4	0.2049	−0.0991	−0.0666	1.0000

FIGURE 11 Screen dumps of calculation of the covariance and correlation matrices in EXCEL. The diagonals of the covariance matrix are the variances of the column, whereas the nondiagonals are the respective covariances. The diagonals of the correlation matrix are 1.

number of observations (177). Accordingly, EXCEL offers two functions *COVARIANCE.P(interval y_i,interval x_i)* (179) and *COVARIANCE.S (interval y_i,interval x_i)* (178) for calculating the covariance of a population and a sample, respectively. Thus, the population covariance is $(n-1)/n$ times the sample covariance. The function in the data analysis operates with the population covariances.

Outliers

Results of a data set that appear to differ unreasonably from the rest are called outliers. It is common practice to calculate regressions and basic statistics with and without suspected outliers and evaluate the results.

A recommended formal test for outliers is the Grubbs' test in which the statistic G is calculated:

$$G = \frac{\text{suspect value} - \bar{x}}{s} \tag{181}$$

where the mean and standard deviation (s) are calculated including the suspect value.

The G is then evaluated using a special table. Formula (182) is applicable if one value is suspected to be an outlier either at the upper or at the lower end of the distribution. Also compare (64). Other formulas are available for more complex situations.

The critical value can be calculated

$$G > \frac{n-1}{\sqrt{n}} \times \sqrt{\frac{t^2_{(\alpha/2, n-2)}}{n - 2 + t^2_{(\alpha/2, n-2)}}} \tag{182}$$

where $t^2_{(\alpha/2, n-2)}$ is the critical value of the t-distribution with $n-2$ degrees of freedom and a significance level of $\alpha/2$. This applies to a two-sided test; for a one-sided use $\alpha/(n-2)$. If the G is larger than the table value or that calculated from Equation (182), then the extreme value is unlikely to have occurred by chance. The table can be found in ISO 5725-2.

The Grubbs' test should not be applied if the number of observations is less than 6 or more than 50.

The G-statistic as described above is equal to the z-score (51) and can be evaluated as such by using a normal cumulative table or the EXCEL: *NORMSDIST(G)*.

Example

A set of observation had a mean of 45.8 and $s(x)$ of 5.1. One observation of 58 was a suspected outlier.

$G = z = \dfrac{58 - 45.8}{5.1} = 2.4$, which is more than the expected $z = 1.96$ for a 97.5 probability of not belonging to the distribution. *Note* that 58 is above the mean and thus a one-sided evaluation. The probability of belonging to the distribution is $1 - NORMDIST(2.4) = 0.08$ (one-sided).

Another test for outliers is the Dixon test (Q-test):

$$Q = \frac{\text{suspect value} - \text{nearest value}}{\text{range of values}} \tag{183}$$

The critical values can be found in a special table. The Dixon test is usually applicable to 3-10 observations. The Grubbs' and Dixon's tests assume a Gaussian distribution of the quantity values.

Rejection of suspect values should be made with great care. Even if also grossly deviating values may belong to the distribution in question, they may have an undue effect on calculated quantities, e.g., mean, standard deviation regression, and correlation. The effect of suspect outliers can be minimized by trimming or winsorizing the data set (64) and (65), respectively, but are of less or no importance in nonparametric calculations.

Leverage

Extreme values and outliers may have an influence (*leverage*) on the regression. An observation far from the centroid (mean of X and Y) is usually a leverage point but not necessarily an influence point. Influence points have an influence on the regression function and have a tendency to "draw the line closer." Leverage points that are not influence points may have a profound effect on the correlation coefficient and the variance of the variables. The larger the leverage, the larger influence it will have on one or several of the properties of the regression.

The leverage of an observation x_j can be expressed quantitatively:

$$h_j = \frac{1}{n} + \frac{(x_j - \bar{x})^2}{\sum_{i=1}^{n}(x_i - \bar{x})^2} = \frac{1}{n} + \frac{(x_j - \bar{x})^2}{(n-1) \times s_x^2} = \frac{1}{n} + \frac{1}{n-1} \times \left(\frac{x_j - \bar{x}}{s_x}\right)^2 \tag{184}$$

where $0 \leq h_j \leq 1$, and n is the number of observations.

The leverage is thus mainly influenced by the distance of the independent variable from the centroid. The last factor in Equation (184) is equal to the squared z-score (51) of the observation. The larger h_j, the more influence it has on the regression. If the observation is in line with a regression line dominated by the bulk of observations (characterized by a small residual), it will have little influence on the regression but on the variance of the distribution of the variables. Often an h_j of 0.9 is used as a cutoff value. Extreme high or low quantity values may have a large influence but still not be regarded as "outliers" if they are found on or close to an otherwise defined regression line.

Incidentally, the variance of the predicted value $\hat{y}$ includes part of the leverage:

$$s_{\hat{y}}^2 = s_{y.x}^2 \times \frac{1}{n-1} \times \left(\frac{x_j - \bar{x}}{s_x}\right) \tag{185}$$

If h_j exceeds

$$h_j > \frac{2 \times p}{n} \quad \text{or} \quad h_j > \frac{3 \times p}{n} \tag{186}$$

where p is the number of predictors (for bivariate linear regression $p = 1$), the point is regarded as a leverage point and needs special consideration.

The number of observations in a regression analysis is crucial. The number of observations is directly included in the calculations of the slope and its uncertainty, the intercept, the correlation coefficient, and the leverage.

COMPARING QUANTITIES

Ideally, a calibration of a measurement procedure with the same calibrator would provide the same result when the same quantity is measured in the same sample. For several reasons, this is not always the case when analyzing biological samples. Therefore, laboratories with many instruments for measuring the same quantities compare the performance of measurement procedures using real samples, usually patient samples.

In clinical research, comparisons of outcomes of studied diagnostic procedures or treatments are important strategies, often with a view to find and confirm the diagnostic usefulness of a marker or to find surrogate markers for complex physiological phenomena. Both questions may be addressed and answered by comparison of results from measurements of patient samples and evaluating the results by regression and correlation studies.

If the same quantities are measured in a method comparison, it is fair to assume that the regression will be linear. If different quantities are measured, e.g., in a calibration of a measurement procedure or comparison of diagnostic procedures, the regression may take any form and be described by for instance linear, logarithmic, exponential, or polynomial functions. It is thus logical to suspect that if the regression is nonlinear and the correlation very poor, the procedures measure different quantities.

Particularly in clinical experiments, the correlation may be poor, often due to imprecision or interfering substances, generally, or at certain concentrations. As demonstrated in Equation (172), high t-values and thus the significance may be obtained in a comparison at a given coefficient of variation (r), simply by increasing the number of observations. This prompts for great care before too long-reaching conclusion can be drawn from a significant r-value.

Graphical Representation

It is usually recommended that a scattergram (Figures 7 and 12 (left)) is first created in a comparison. This is to give the scientist a broad overview of the distribution of results and spot possible outliers. Usually, the "equal line" and some regression function are also presented.

To facilitate the evaluation of a comparison, "difference graphs" (Figure 12, right) are also usually constructed and often demanded for publication in scientific journals. The

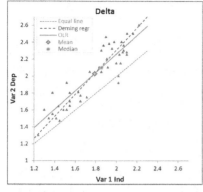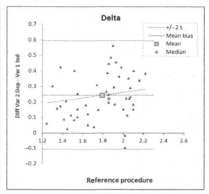

FIGURE 12   Regression and difference graphs. The Deming and OLR functions are shown, $Y=1.30X-0.30$ and $Y=1.09X+0.08$, respectively. Equal variances ($\lambda_i=1$) for the methods were assumed for the Deming regression. Both regressions are centered on the average of the dependent and independent variables, and therefore, the regression lines cross at that point. The difference between the mean and median is enhanced in the difference graph. *Note*: The scales are equal. In the difference graph the independent variable was chosen as reference. The X-axis represents the equal line in the scatterplot and the regression function is $Y=0.09X+0.08$ (see below). The mean of the differences (bias) and $\pm2\ s$ are shown. The correlation coefficients for the scatter and difference data were 0.748 and 0.019, respectively, illustrating the gain in "resolution" of the differences.

difference graph displays the difference between the measurements plotted against the mean of the results or, if the comparative method (independent variable) can be regarded as a reference measurement procedure, against these values (Var 1) directly. Difference plots with the mean of the variables as the independent variable are known as Bland-Altman graphs.

The mechanics behind the design of the difference graph can be understood as subtracting $Y=X$ from the regression $Y=bX+a$, i.e., forming a new function where Y represents the difference and X still represents the comparative method:

$$Y - Y_1 = b \times X - X_1 + a;$$
$$Y = X \times (b - 1) + a \tag{187}$$

As seen from this formula, the regression function of the differences will have a slope which is $45°$ less than that of the original observations. In other words, the data of difference graph appear tilted 45° clockwise in relation to the original regression function. Consequently, the equal line of the original will be represented by the X-axis and the slope of the regression function of the difference graph will be 1 (tan(45°)) less than that of the original data, whereas the intercept (a) is unchanged (Figure 12). The correlation coefficient will be decreased in comparison with that of the original and thus differences appear enhanced. There is no unique new information in the difference graph but a visual enhancement.

If the differences are normally distributed, the mean difference and its standard deviation can be calculated and displayed in the difference graph (Figure 12) and compared with target values.

Typical questions that are answered by the difference graph, in addition to mean and dispersion, are if the difference increases or decreases with the concentration and if the dispersion seems to be constant or change with concentration.

A complementary graph (Figure 13), based on the cumulative empirical distribution function for the differences, may be

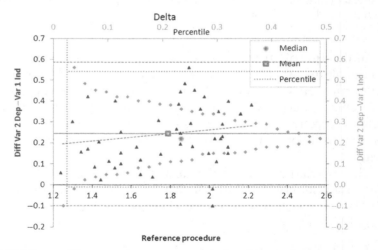

FIGURE 13 The cumulative, empirical distribution function superimposed on the difference graph. The peak of the tilted mountain coincides with the median of the differences. The dotted vertical and horizontal line corresponds to the 2.5 and 97.5 percentiles, i.e., the central 95 % of the observations, centered round the median. Compare with Figure 5.

superimposed on the difference graph to better illustrate the distribution of the differences. The emperical distribution function will coincide with the sigmoid cumulative curve by a normal distribution if the number of observations is sufficiently large. To create an emperical distribution functions the differences are ranked, any ties resolved and the percentiles of the ranks calculated. This produces a sigmoid which will, however, be liable to effects of a limited number of observations. The obtained distribution can be compared with what would be expected from a normal distribution which would be just another application of the Q-Q plot.

The function is then mirrored around the 50 % percentile (the median) which results in a mountain-shaped curve. This can then be tilted 90° clockwise and superimposed on the difference graph (Figure 13).

PERFORMANCE CHARACTERISTICS

Definitions

In a dichotomous decision situation, i.e., only two alternatives as deciding if a person has a given disease or condition or not, there are four possible outcomes. These can be defined in a 2×2 frequency table in which the number of individuals belonging to categories (healthy and nonhealthy, respectively) of an independent classification is in one row. The number of individuals or items that are tested positive and negative, respectively, is reported in columns (Table 14 and Figure 14).

Concordance between diagnostic test results and the independently found diagnosis or property are recorded as true positive = TP, true negative = TN, false positive = FP, and false negative = FN.

The performance can be expressed as diagnostic (nosographic) sensitivity, specificity, predictive value of a positive result = PV(+) and predictive value of a negative result = PV(−):

$$Diagnostic\ Sensitivity\ (Sens) = \frac{TP}{TP + FN} \qquad (188)$$

TABLE 14 Contingency Table or 2×2 Table for Classification of Dichotomous Results Assuming a Positive Test Result for Condition 1 and Negative for Condition 2

	Negative outcome of test	Positive outcome of test	
Condition 1	False negative (FN)	True positive (TP)	$Sensitivity = PV(-) = \frac{TP}{TP+FN}$
Condition 2	True negative (TN)	False positive (FP)	$Specificity = PV(-) = \frac{TN}{TN+FP}$
	$PV(-) = \frac{TN}{TN+FN}$	$PV(-) = \frac{TP}{TP+FP}$	$Efficiency = \frac{TP+TN}{TP+TN+FP+FN}$

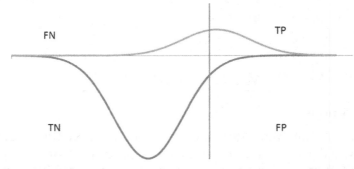

FIGURE 14 Illustration of the contingency table in Table 14. The condition 1 (e.g., a diseased group) is above the X-axis and the condition 2 (e.g., a nondiseased group) below. The vertical line is the "cutoff" of a diagnostic marker. The frequency distributions are idealized, particularly the group of diseased would be skewed to the right.

$$Diagnostic\ Specificity\ (Spec) = \frac{TN}{TN + FP} \qquad (189)$$

The sensitivity and specificity is changed by changing the cutoff between categories (e.g., considered healthy and non-healthy), reference value or decision value. In the table, this would imply changing the relation between the number of items in the columns. An increase in one of the quantities will invariably cause a decrease in the other (cumulative distribution analysis, CDA test, Figure 16).

$$PV\ (+) = \frac{TP}{TP + FP} \qquad (190)$$

$$PV\left(-\right) = \frac{TN}{TN + FN} \tag{191}$$

$$Efficiency = \frac{TP + TN}{TP + TN + FP + FN} =$$

$$Sens \times Prev + Spec \times \left(1 - Prev\right) =$$

$$Prev \times \left(Sens - Spec\right) + Spec \tag{192}$$

The *Efficiency* is thus directly proportional to the *Prevalence of disease.*

Efficiency is also known as "Index of validity," "Agreement" or *Accuracy.*

"Index of agreement" is *defined* as kappa (κ) (237)

Prevalence of disease (*pre-test probability; Prev*)

$$= \frac{TP + FN}{TP + TN + FP + FN} \tag{193}$$

Note Expressions (190)–(192) depend on the prevalence of disease (193), whereas Equations (188) and (189) are characteristics of a diagnostic procedure that is used for a specific purpose, using defined discriminators (e.g., cutoffs). They may thus be regarded as constants in that context.

Bayes' Theorem

The theorem describes a method to estimate the post-test probability by applying key characteristics of an investigation to the pretest probability.

$$Likelihood\ ratio\left(+\right)\left(LR(+)\right) = \frac{Sensitivity}{1 - Specificity} \tag{194}$$

$$Likelihood\ ratio\left(-\right)\left(LR(-)\right) = \frac{1 - Sensitivity}{Specificity} \tag{195}$$

The LR(+) and LR(−) are also known as Bayes' factors.

$$Odds = \frac{Probability}{1 - Probability} \tag{196}$$

$$Probability\,(Prob) = \frac{Odds}{1 + Odds} \tag{197}$$

Pre-test probability = *Prevalence of disease*, see Equation (193)

$$Pre\text{-}test\;odds = \frac{Prevalence}{1 - Prevalence} \tag{198}$$

$$Post\text{-}test\;odds = pre\text{-}test\;odds \times LR \tag{199}$$

This relation summarizes the Bayes' theorem.

$$Post\text{-}test\;probability\;[PV\,(+)\,or\;PV\,(-)] = \frac{Post\text{-}test\;odds}{1 + Post\text{-}test\;odds}$$
$$\tag{200}$$

Prob of disease if a post-test $(PV(+))$:

$$\frac{Sens \times Prev}{Sens \times Prev + (1 - Spec) \times (1 - Prev)} \tag{201}$$

Prob of no disease if a neg test $(PV(-))$:

$$\frac{Spec \times (1 - Prev)}{Spec \times (1 - Prev) + (1 - Sens) \times Prev} \tag{202}$$

Note The post-test probability for a positive and negative result is directly given in the 2×2 table as PV(+) (190) and PV(−) (191), respectively.

The post-test probability can be estimated from the prevalence of disease, sensitivity, and specificity as shown in formulas (201) and (202). A stepwise procedure would require to first calculate the LR(+) and LR(−), the pre-test odds and the posttest odds. A post-test probability is then obtained from Equation (200) for PV(+) and [1 − (200)] for PV(−).

An online calculator for quantities related to Bayesian logics is available at http://araw.mede.uic.edu/cgi-bin/testcalc.pl.

The relation between the pre-test probability and the post-test probability can be visually displayed in the Fagan nomogram (Figure 15) in which information on the pre-test

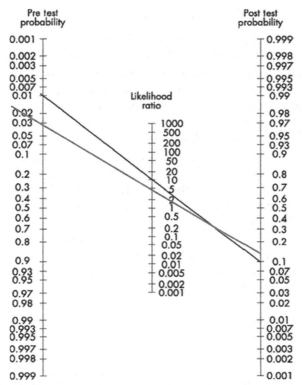

FIGURE 15 Fagan nomogram. Draw a straight line between the pretest probability and the LR(+) and read the posttest probability on the right-hand scale.

probability (prevalence of disease) and the likelihood ratio will indicate the post-test probability and thus the gain by performing the test.

A LR(+) of 1 indicates that there is no gain in performing the test. Often an LR(+) of at least three is required in clinical work.

Risk ratio = relative risk (RR):

$$RR = \frac{I_e}{I_{ne}} \tag{203}$$

where I_e is the incidence of exposed individuals and I_{ne} incidence of not exposed individuals.

Receiver Operating Characteristics

The *diagnostic sensitivity* (*Y*) plotted against (1 − *diagnostic specificity*) (*X*) for many chosen cutoff or reference values is the receiver operating characteristics (ROC) curve (Figure 16).

This is the same as plotting (1 − β) as the dependent variable (*Y*) against α as the independent variable (*X*).

The *ROC curve* is thus a graphical representation of the trade-off between the false-negative and false-positive rates for every possible quantity value. Equivalently, the *ROC curve* is the representation of the trade-offs between sensitivity and specificity. By tradition, the plot shows the false-positive rate (α) (1 − *specificity*) on the *X*-axis and (1 − the false-negative rate), i.e., (1 − β) (*sensitivity*) on the *Y*-axis.

The *ROC curve* also represents the likelihood ratio (LR+) (194) for each tested cutoff value.

The sensitivity and specificity concepts as defined above are most suited for binary or dichotomous situations, i.e., a "yes" or "no" answer. This limitation spills over on the *ROC curve*. A limitation of the ROC curve is its negligence of the prevalence of the condition.

Youden index.

If *sensitivity* and *specificity* are equally important, the Youden index [*J*] will indicate the performance (the higher the better) at a given cutoff. *J* is sometimes used to define an optimal cutoff (*c*).

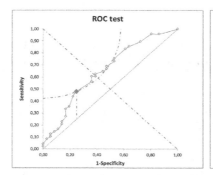 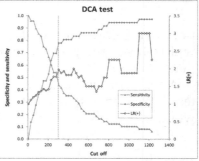

FIGURE 16 ROC and CDA curves. In the ROC curve, the quarter circle represents the *K*-index and the vertical line the *J*-index. The *sensitivity = specificity* line and its perpendicular; the theoretical optimum. The vertical line in the CDA plot shows the *sensitivity, specificity,* and *LR(+)* at the chosen cutoff.

$$J = Max_c \left(sensitivity_c + specificity_c - 1 \right) \tag{204}$$

The maximum value of the *Youden index* is 1 (perfect test) and the minimum is 0 when the test has no diagnostic value. The value of one [1] will be achieved when the test and comparative populations are completely separated (see figure 14).

The optimal outcome would be when the *sensitivity* is equal to 1, and at the same time, the *specificity* also equals 1, i.e., the false-positive rate (1 −*specificity*) is zero (0). The point representing this combination will be in the upper left corner of the graph. The closer a *ROC* curve is to this ideal situation, the better the marker performs, given that *sensitivity* and *specificity* are of equal diagnostic importance. This is another way of expressing the Youden index (204).

The ROC curve is a summary of the information, and as such, some information is lost, particularly the value of each cutoff. The *CDA* (Figure 16, right) displays the *sensitivity* and *specificity* against the cutoff values on the X-axis, which is another compromise which addresses this property. The CDA is thus a more useful tool to choose and describe the effect of a particular cutoff. It also demonstrates the influence of changing the cutoff value.

Example

B-Glucose concentrations were measured in 200 patients aged 40-60 years. In this age group, the prevalence of disease was estimated to 6 % by an independent method. The specificity was estimated to 0.85 and the sensitivity 0.95. The diagram below (Figure 17) illustrates the possible outcome, increasing the pre-test probability of 6 % to a post-test probability of 29 %. The $PV(-)$ is about 1 and thus the test is useful for rule-out at this prevalence.

Area Under the Curve

The area under the *ROC curve* (*AUC*) summarizes the performance. If the sum of the *sensitivity* and the *specificity* equals one (*TP*=*FP*), i.e., the area under the curve (AUC)=0.5 and the *ROC curve* follows the diagonal, then the performance is

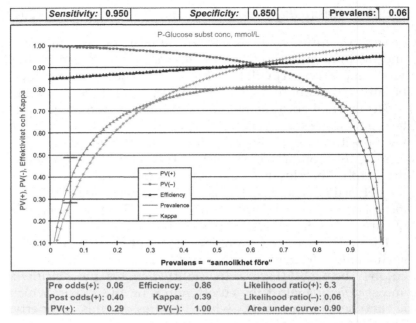

| Sensitivity: | 0.950 | | | Specificity: | 0.850 | | Prevalens: | 0.06 |

Pre odds(+):	0.06	Efficiency:	0.86	Likelihood ratio(+): 6.3
Post odds(+):	0.40	Kappa:	0.39	Likelihood ratio(−): 0.06
PV(+):	0.29	PV(−):	1.00	Area under curve: 0.90

FIGURE 17 Simultaneous display of the relation between diagnostic performance characteristics and the prevalence of disease. Imagine the violet vertical line moving horizontally and read the outcome.

no better than chance. Compare LR(+) equal to one [1] in the Fagan diagram, Figure 15.

An approximate value of *AUC* can be estimated by adding the area of trapeziums (a rectangle with a triangle on top) formed by connecting consecutive points. Thus, if $(1 - specificity)$ of two adjacent observations is X_1 and X_2 and the corresponding sensitivity Y_1 and Y_2 for the points limiting the trapezium, then its area (A_{1-2}) is

$$A_{1-2} = Y_1 \times (X_1 - X_2) + \frac{(Y_2 - Y_1) \times (X_1 - X_2)}{2} = \frac{(X_1 - X_2) \times (Y_1 + Y_2)}{2} \tag{205}$$

The AUC is obtained by adding the individual $A(x_i - x_{(i+1)})$. The more trapeziums that are identified and defined (i.e., the smaller the difference $X_1 - X_2$), the better is the estimate. It should be recognized that at lower specificities, an undue contribution to the area is made by the area of the trapeziums under the equal line. AUC calculation is offered in many software packages.

Reference Values

The information derived from the *ROC* curve can also be used to select a reference value or decision value depending on the priority given to the *sensitivity* or *specificity* in a given clinical setting. Since the theoretical maximal efficiency occurs when *sensitivity* = *specificity* = 1, i.e., the upper left corner of the *ROC curve*, the cutoff corresponding to a minimized distance *d* (K-index) between the potential reference value and the corner would be an optimal compromise:

$$d = \sqrt{(1 - sens)^2 + (1 - spec)^2} \qquad (206)$$

In a *ROC curve*, the *d* will be represented by a quarter of a circle since there are many solutions to Pythagoras' theorem with only *d* defined (206) (see Figure 16).

A high *specificity* will tend to rule out disease in a decision situation and a high *sensitivity* will rule in, but the outcome is also influenced by the prevalence of disease (see Figure 17).

A high *sensitivity* (rule in) is often preferred for screening purposes, if followed by a procedure with high *specificity*.

ESTIMATION OF MINIMAL SAMPLE SIZE (POWER ANALYSIS)

Error Types

To determine the power of a test and the minimal sample size, one has to consider the null hypothesis, the probability of rejecting false-positive results (α), the probability of not rejecting false-negative results (β), and the dispersion of the sample results and the assumed difference between results and whether a one- or two-tailed statistical analysis is planned.

Thus, the null hypothesis can either be true or false and we can make two types of error; if rejected when it is true, i.e., the type I or α-error and if not rejected when false, i.e., the type II or β-error.

The type II error, β-error, is when the difference overlaps with the values that would not be regarded as a difference. This can only be one-sided.

The p-value of a significance test equals the probability that a result occurs, i.e., more extreme than those when the null hypothesis is true. For instance, if the null hypothesis is that two values are the same, a high p-value will be supportive, whereas a small (usually $p < 0.05$) will not and the values are regarded as different with a 95 % probability (two-sided). And there is a 5 % probability ($\alpha = 0.05$) that we make a mistake in this judgment.

Consider two populations which averages differ by d and which overlap to a certain degree. The null hypothesis states that there is no difference between the populations, i.e., $d = 0$. However, a tail of one of the distributions (α) coincides with the other population and a tail of this (β) coincides with the first. The tails are defined by a "cutoff." Therefore, if one increases, the other will decrease and the analogy with the diagnostic specificity ($1 - \alpha$) and diagnostic sensitivity ($1 - \beta$) is obvious (see above):

$$Specificity = 1 - \alpha \tag{207}$$

and

$$Sensitivity = 1 - \beta \tag{208}$$

where α is the probability of rejecting a false positive and β is the probability of *not* rejecting a false positive, respectively. That is, α is the false-positive rate and β is false-negative rate, usually linked to discussions on the probability of identifying differences (see below).

Power of a Test

Power is defined as the probability that a statistical test will reject the null hypothesis when it is false. Therefore, power is equal to $1 - \beta$ or sensitivity. A commonly accepted power $= 0.80$, i.e., $\beta = 0.20$.

$$\alpha = \textit{"probability of falsely accepting the alternate hypothesis,"}$$
$$\textit{i.e., rejecting the null hypothesis when true} \tag{209}$$

$\beta =$ "probability of falsely accepting a null hypothesis,"

i.e., not rejecting the null hypothesis when it is false   (210)

$1 - \beta =$ "power"   (211)

The acceptable size of the ratio β / α is conditional to the purpose of the power analysis; if false positives are critical, increase the ratio, e.g., by decreasing the α, if false negatives are more important, reduce the ratio. An acceptable ratio in clinical practice is often set to $\dfrac{0.20}{0.05} = 4$.

Sample Size

If $\bar{X} - t \times \dfrac{s}{\sqrt{n}} < \mu < \bar{X} + t \times \dfrac{s}{\sqrt{n}}$, then the CI or margin of error (E) is

$$CI = t \times \frac{s}{\sqrt{n}}$$ (212)

and

$$n = \left(t \times \frac{s}{CI} \right)^2$$ (213)

The same is valid for proportions (p). If the proportion is not known, use $p = 0.5$ that has the highest standard deviation of all proportions.

There are other rules to estimate the sample size. If we want to estimate the sample size from the acceptable standard error of the mean $s(\bar{x})$, this formula can be used:

$$n \approx z_\alpha^2 \times \left(\frac{u(x)}{s(\bar{x})} \right)^2$$ (214)

Example

Suppose the method uncertainty ($u(x)$) is 20 and the acceptable standard error of the mean $s(\bar{x}) = 10$, then a reasonable number of observations is 16 with 95 % confidence

$(\alpha = 0.05; z = 2)$ $n \approx z_\alpha^2 \times \left(\dfrac{u(x)}{s(\bar{x})} \right)^2 = 2^2 \times \left(\dfrac{20}{10} \right)^2 = 16$.

It is argued slightly different to estimate the necessary number of observations to identify a difference between two results (x_1 and x_2).

Example

Assume the same number of observations in the two groups (n) and the same variance $(s(x))^2$ of the results. This leads toward a Student's independent t-test. If the difference of the results of measurements is expressed in standard deviations (cf. z-score (51)), the expression is

$$d = \frac{x_1 - x_2}{s} \tag{215}$$

The necessary number of observations in each sample is

$$n = \frac{2 \times \left(z_{(1-\alpha/2)} + z_{(1-\beta)}\right)^2}{\left(\frac{x_1 - x_2}{s(x)}\right)^2} = \frac{2 \times s(x)^2 \times \left(z_{(1-\alpha/2)} + z_{(1-\beta)}\right)^2}{(x_1 - x_2)^2} =$$

$$\frac{2 \times \left(z_{(1-\alpha/2)} + z_{(1-\beta)}\right)^2}{d^2} \tag{216}$$

If $\alpha = 0.05$ and $\beta = 0.2$, i.e., the $\beta/\alpha = 4$ (see above), $z_{(1-\alpha/2)}$ and $z_{(1-\beta)}$ are 1.96 (in EXCEL: $NORMSINV(1 - \alpha/2)$) and 0.84 ($NORMSINV(1 - \beta)$), respectively. The numerator will be 15.7 which is rounded to 16 and a simplified formula will be

$$n = \frac{16}{\left(\frac{x_1 - x_2}{s(x)}\right)^2} = \frac{(s(x))^2 \times 16}{(x_1 - x_2)^2} = \frac{16}{d^2} = \frac{16}{(z - score)^2} \tag{217}$$

where x_1 and x_2 are the sample means, $s(x)$ is the standard deviation of both samples, and n the number of observations in each sample. The factor 16 refers to a two-sample situation; in a one-way the factor will be 8.

The detectable difference can be calculated from Equation (214), if inverted:

$$d^2 = \frac{16}{n}; \quad d = \frac{4}{\sqrt{n}} \tag{218}$$

This is the difference that can be detected with a 95 % confidence for a given sample size.

If there are only two observations the MD will be

$$d \approx \frac{4}{\sqrt{2}} \approx 2.83 = \frac{x_1 - x_2}{s_x}; \quad x_1 - x_2 = 2.83 \times s_x \tag{219}$$

Compare the combined uncertainty of the difference between two observations with the same uncertainty $u(A) = u(C) = z_{1-\alpha/2} \times u(A) \times \sqrt{2} \approx 2.77 \times u(a)$ (cf. MD (80)). This should be exceeded to indicate a significant difference between the results. The factor 4 is an approximation and represents $2 \times (1.96 + 0.84) = 15.68$, changing the factor in Equation (219) to 2.77.

Sample Size If Given the %CV

If the %CV is the same for both methods, the relative difference is first estimated as

$$RD = \frac{x_1 - x_2}{\frac{(x_1 + x_2)}{2}} \tag{220}$$

and the number of samples in each group

$$n \approx \frac{16 \times CV^2}{RD^2} \tag{221}$$

Example

If the relative difference to be detected is, e.g., 20 % and the relative uncertainty (coefficient of variation: %CV) 30 %, the following is obtained:

$$n \approx \frac{16 \times \%CV^2}{RD^2} = \frac{16 \times 0.3^2}{\left(\frac{1-0.8}{(1+0.8)}\right)^2} = \frac{1.44}{0.049} \approx 29$$

If the relevant difference to be detected is 10 %, the number of observations needs to be about (130). If the comparison is with a standard, i.e., only one group is needed, then the factor in the nominator is 8.

If the averages are known (x_1 and x_2) and the %CV specified, then Equation (221) can be written as

$$n \approx \frac{16 \times (CV)^2}{(ln(x_1) - ln(x_2))^2} \tag{222}$$

Number Needed to Treat or Harm

Define EE and CE as the number of Events in the Experimental and Control groups, respectively, and EN and CN the number of nonevents in the Experimental and Control groups, respectively. Let ES and CS be the number of subjects in the Experimental group and Control groups, respectively. The EER is the Experimental Event Rate and CER as Control Event Rate; then

$$ES = EE + EN; \quad CS = CE + CN; \quad EER = \frac{EN}{ES} \text{ and } CER = \frac{CN}{CS} \tag{223}$$

Relative risk (risk ratio cf. 203):

$$RR = \frac{EER}{CER} \tag{224}$$

and

Experimental event odds:

$$EEO = \frac{EE}{EN} \tag{225}$$

Control event odds:

$$CEO = \frac{CE}{CN} \tag{226}$$

Odds ratio:

$$OR = \frac{EEO}{CEO} = \frac{\frac{EE}{EN}}{\frac{CE}{CN}} = \frac{EE \times CN}{CE \times EN} = \frac{TP \times TN}{FN \times FP} \qquad (227)$$

Efficacy of treatment $= 1 - RR$ $\qquad$ (228)

Number needed to treat or harm:

$$\frac{1}{EER - CER} \qquad (229)$$

if $(229) < 0$ then NNT (Number Needed to Treat);
if $(229) > 0$ then NNH (Number Needed to Harm)
The larger the absolute value of the NNT, the less efficient is the treatment.

AGREEMENT BETWEEN CATEGORICAL ASSESSMENTS (KAPPA (κ)-STATISTICS)

This problem is faced when two measurement procedures which report results on an ordinal scale are compared and the number of agreeing results can be organized in a cross-table (contingency table, frequency table). It is also used when observers are categorizing patients or events into two or several groups, e.g., two or more experts evaluating test results. A special case is the 2×2 frequency table where only two groups are considered (see below).

Any number of categories (groups) can be studied. The table is characterized by the same number of rows and columns.

Example
Enter the number of observations in each group for each observer in the appropriate cells (Table 15):
Proportion of agreement by chance (expected):

$$P_C = (11 \times 21 + 12 \times 22 + 13 \times 23 + 14 \times 24)/(N)^2 \qquad (230)$$

Proportion of observed agreement:
$$P_O = (A + F + K + Q)/N \qquad (231)$$

TABLE 15   Notation in the Example

		Method (Observer) 1				
		Group 11	Group 12	Group 13	Group 14	Total
Method (observer) 2	Group 21	A	B	C	D	21
	Group 22	E	F	G	H	22
	Group 23	I	J	K	L	23
	Group 24	M	O	P	Q	24
	Total	11	12	13	14	Total (N)

$$Kappa\ (\kappa) = \frac{P_O - P_C}{1 - P_C} = 1 - \frac{1 - P_O}{1 - P_C} \tag{232}$$

Note The κ-value is influenced by the *bias* between the observers defined as the difference between Method (Observer) A and Method (Observer) B in their assessment of the frequency of occurrence of a condition. A high value of the κ-value indicates a high degree of concordance between observers.

In a 2×2 table, the bias can be quantified as the Bias IndexBI (symbols in the example are retained, assuming that all input cells except A, B, E, and F are zero [0]):

$$BI = \frac{A + B}{N} - \frac{A + E}{N} = \frac{B - E}{N} \tag{233}$$

thus reflecting the difference between cells of disagreement B and E.

BI results can take values between (-1) and $(+1)$.

Kappa (κ) can be corrected for the bias BAK:

$$BAK = 1 - \frac{1 - P_O}{1 - \frac{(11+21)^2 + (12+22)^2}{4 \times N^2}} \tag{234}$$

The value of κ is also affected by the relative probabilities of the "Yes" and "No" answers (in a 2×2 table). This is called *Prevalence Index*PI:

$$PI = \frac{A - F}{N} \tag{235}$$

and is the difference between cells of agreement A and F. PI ranges from (-1) to $(+1)$.

A *kappa* (κ) *value* that is both prevalence and bias adjusted is PABAK:

$$PABAK = 2 \times P_O - 1 \tag{236}$$

PABAK ranges from (-1) *to* $(+1)$ like *kappa*(κ), but the interpretation may be different from that of the uncorrected *kappa* (κ) *value*.

The value of κ is depending on all these indexes:

$$\kappa = \frac{PABAK - PI^2 + BI^2}{1 - PI^2 + BI^2} \tag{237}$$

Clearly, if either of BI or PI takes on extreme values, the interpretation of the κ-*value* is difficult.

κ can take any value between -1 and $+1$, $\kappa = 1$ is total agreement, $\kappa = 0$ (agreement expected to chance); $\kappa < 0$ indicates less than expected (rare), total disagreement at $\kappa = (-1)$:

$$se(p) = \pm \sqrt{\frac{P_O \times (1 - P_O)}{n \times (1 - P_C)^2}} \tag{238}$$

Agreement in a 2×2 Table

The efficiency of a diagnostic test is described in a 2×2 table by the *sensitivity* (188) and *specificity* (189) of the test. The *efficiency* is the sum of true results relative to all observations (192). Considering the influence of chance on the efficiency gives an expected "efficiency":

$$p_e = \frac{(TP + FN) \times (TP + FP) + (TN + FN) \times (TN + FP)}{N^2} \tag{239}$$

where N is the total number of observations.

The index of agreement, i.e., kappa is

TABLE 16 Evaluation of κ-Values

Kappa (κ)	Agreement
0.00	Poor
0.01-0.20	Slight
0.21-0.40	Fair
0.41-0.60	Moderate
0.61-0.80	Substantial
0.81-1.00	Almost perfect

The agreement between test and diagnosis.

$$\kappa = \frac{efficiency - p_e}{1 - p_e} \tag{240}$$

κ is interpreted as the difference between the found efficiency and the expected relative to that possible, considering the chance.

κ has been calculated and included in Figure 17. It is noteworthy how chance decreases the efficiency of a test at high and low prevalence of disease. This is particularly important in validating a diagnostic marker, i.e., evaluating it being fit for purpose (Table 16).

The approximate standard error of κ is

$$se(\kappa) = \sqrt{\frac{Efficiency \times (1 - Efficiency)}{n \times (1 - p_e)^2}} \tag{241}$$

Example

Compare two hypothetical 2×2 tables, A and B

	Test				Test		
A	**Positive**	**Negative**	**Sum**	**B**	**Positive**	**Negative**	**Sum**
Diseased	10	5	15	Diseased	20	5	25
Healthy	5	80	85	Healthy	5	70	75
Sum	15	85	100	Sum	25	75	100

Sensitivity	0.67	Sensitivity	0.8
Specificity	0.94	Specificity	0.93
Efficiency	0.9	Efficiency	0.9
Expected efficiency	0.75	Expected efficiency	0.63
Kappa (κ)	0.61	Kappa (κ)	0.73

Although the *efficiency* is the same, the κ-value indicates a better agreement between test and diagnosis in the B example, which is also demonstrated in the *sensitivity*, this time.

Some Metrological Concepts*

METROLOGY, ACCURACY, TRUENESS, AND PRECISION

quantity
property of a phenomenon body, or substance, where the property has a magnitude that can be expressed as a number and a reference.

Note 1 A reference can be a measurement unit, a measurement procedure, a reference material, or a combination of such.

Note 2 The preferred IUPAC-IFCC format for designations of quantities in laboratory medicine is "System—Component; kind of quantity."

kind of quantity
aspect common to mutually comparable quantities.

Note 1 The division of "quantity" according to "kind of quantity" is to some extent arbitrary.

*Extracted from JCGM 200 2012 (VIM 2012): International Vocabulary of Metrology—Basic and General Concepts and Associated Terms (VIM), 3rd edition, 2008 version with minor corrections.

This important source of defined terminology is freely downloadable from http://www.bipm.org/utils/common/documents/jcgm/JCGM_200_2012.pdf.

The terms we preset here are not necessarily presented in the alphabetical order. The choice of terms is limited and the reader is advised to consult with the original document. Notes and examples may not be cited *in extenso*.

Example 1

The quantities of diameter, circumference, and wavelength are generally considered to be quantities of the same kind, namely of the kind of quantity called length.

Example 2

The quantities heat, kinetic energy, and potential energy are generally considered to be quantities of the same kind, namely of the kind of quantity called energy.

Note 2 Quantities of the same kind within a given system of quantities have the same quantity dimension. However, quantities of the same dimension are not necessarily of the same kind.

quantity value

value of a quantity, value.

number and reference, together expressing magnitude of a quantity.

Example 1 Length of a given rod: 5.34 m or 534 cm.

Example 2 Mass of a given body: 0.152 kg or 152 g.

Example 3 Celsius temperature of a given sample: −5 °C.

Example 4 Molality of Pb^{2+} in a given sample of water: 1.76 μmol/kg.

Example 5 Arbitrary amount-of-substance concentration of lutropin in a given sample of human blood plasma (WHO International Standard 80/552 used as a calibrator): 5.0 IU/L, where "IU" stands for "WHO International Unit"

Note 1 According to the type of reference, a quantity value is either a product of a number or a measurement unit (see Examples 1–4).

The measurement unit one is generally not indicated for quantities of dimension one or

> a number and a reference to a measurement procedure or
> a number and a reference material (see Example 5).

measurement

process of experimentally obtaining one or more quantity values that can reasonably be attributed to a quantity.

Note 1 Measurement does not apply to nominal properties.

Note 2 Measurement implies comparison of quantities and includes counting of entities.

Note 3 Measurement presupposes a description of the quantity commensurate with the intended use of a measurement result, a measurement procedure, and a calibrated measuring system operating according to the specified measurement procedure, including the measurement conditions.

measuring system

set of one or more measuring instruments and often other devices, including any reagent and supply, assembled and adapted to give information used to generate measured quantity values within specified intervals for quantities of specified kinds.

Note

A measuring system may consist of only one measuring instrument.

measuring instrument

device used for making measurements, alone or in conjunction with one or more supplementary devices.

Note 1 A measuring instrument that can be used alone is a measuring system.

Note 2 A measuring instrument may be an indicating measuring instrument or a material measure.

measurand

quantity intended to be measured.

Note 1 The specification of a measurand requires knowledge of the *kind of quantity*, substance carrying the quantity, including any relevant component, and the chemical entities involved.

Note 2 The measurement, including the measuring system and the conditions under which the measurement is carried out, might change the phenomenon, body, or substance such that the quantity being measured may differ from the measurand as defined.

Example 1

The length of a steel rod in equilibrium with the ambient Celsius temperature of 23 °C will be different from the length at the specified temperature of 20 °C, which is the measurand. In this case, a correction is necessary.

Note 1 In chemistry, "analyte," or the name of a substance or compound, is a term sometimes used for "measurand." This usage is erroneous because this term does not refer to quantity.

Note 2 The measurand is a *quantity*, e.g., "glucose concentration in serum" where glucose is the *analyte* or *component* and serum is the *system* or *matrix.*

ordinal quantity

quantity, defined by a conventional measurementprocedure, for which a total ordering relation can be established, according to magnitude, with other quantities of the same kind, but for which no algebraic operations among those quantities exist.

Example 1 Octane number for petroleum fuel.

Example 2 Subjective level of abdominal pain on a scale from 0 to 5.

Note 1 Ordinal quantities can enter into empirical relations only and have neither measurement units nor quantity dimensions. Differences and ratios of ordinal quantities have no physical meaning.

Note 2 Ordinal quantities are arranged according to ordinal quantity-value scales.

accuracy of measurement

measurement accuracy.

accuracy closeness of agreement between a measured quantity value and a true quantity value of a measurand.

Note 1 The concept "measurement accuracy" is not a quantity and is not given a numerical quantity value. A measurement is said to be more accurate when it offers a smaller measurement error.

Note 2 The term "measurement accuracy" should not be used for measurement trueness, and the term "measurement precision" should not be used for "measurement accuracy," which, however, is related to both these concepts.

Note 3 "Measurement accuracy" is sometimes understood as closeness of agreement between measured quantity values that are being attributed to the measurand.

trueness
measurement trueness.
trueness of measurement.
closeness of agreement between the average of an infinite number of replicate measured quantity values and a reference quantity value.

Note 1 Measurement trueness is not a quantity and thus cannot be expressed numerically, but measures for closeness of agreement are given in ISO 5725.

Note 2 Measurement trueness is inversely related to systematic measurement error, but is not related to random measurement error.

Note 3 Measurement accuracy should not be used for "measurement trueness" and vice versa.

bias
measurement bias.
estimate of a systematic measurement error.

precision
measurement precision.
closeness of agreement between indications or measured quantity values obtained by replicate measurements on the same or similar objects under specified conditions.

Note 1 Measurement precision is usually expressed numerically by measures of imprecision, such as standard deviation, variance, or coefficient of variation under the specified conditions of measurement.

Note 2 The "specified conditions" can be, for example, repeatability conditions of measurement, intermediate precision conditions of measurement, or reproducibility conditions of measurement (see ISO 5725-3:1994).

Note 3 Measurement precision is used to define measurement repeatability, intermediate measurement precision, and measurement reproducibility.

Note 4 Sometimes "measurement precision" is erroneously used to mean measurement accuracy.

intermediate measurement precision
intermediate precision.

measurement precision under a set of intermediate precision conditions of measurement.

intermediate precision condition of measurement
intermediate precision condition.

condition of measurement, out of a set of conditions that includes the same measurement procedure, same location, and replicate measurements on the same or similar objects over an extended period of time, but may include other conditions involving changes.

Note 1 The changes can include new calibrations, calibrators, operators, and measuring systems.

Note 2 A specification for the conditions should contain the conditions changed and unchanged, to the extent practical.

Note 3 In chemistry, the term "interserial precision condition of measurement" or "between series imprecision" is sometimes used to designate this concept.

repeatability
measurement repeatability.

measurement precision under a set of repeatability conditions of measurement.

repeatability condition of measurement
repeatability condition.

condition of measurement, out of a set of conditions that includes the same measurement procedure, same operators, same measuring system, same operating conditions and same location, and replicate measurements on the same or similar objects over a short period of time.

Note 1 A condition of measurement is a repeatability condition only with respect to a specified set of repeatability conditions.

Note 2 In chemistry, the term "intraserial precision condition of measurement" or "within series imprecision" is sometimes used to designate this concept.

reproducibility
measurement reproducibility.

measurement precision under reproducibility conditions of measurement.

reproducibility condition of measurement
reproducibility condition.

condition of measurement, out of a set of conditions that includes different locations, operators, measuring systems, and replicate measurements on the same or similar objects.

Note 1 The different measuring systems may use different measurement procedures.

Notes 2 A specification should give the conditions changed and unchanged, to the extent practical.

UNCERTAINTY CONCEPT AND UNCERTAINTY BUDGET

uncertainty
uncertainty of measurement.

measurement uncertainty.

nonnegative parameter characterizing the dispersion of the quantity values being attributed to a measurand, based on the information used.

Note 1 Measurement uncertainty includes components arising from systematic effects, such as components associated with corrections and the assigned quantity values of measurement standards, as well as the definitional uncertainty. Sometimes estimated systematic effects are not corrected for but, instead, associated measurement uncertainty components are incorporated.

Note 2 The parameter may be, for example, a standard deviation called standard measurement uncertainty (or a specified multiple of it), or the half-width of an interval, having a stated coverage probability.

Note 3 Measurement uncertainty comprises, in general, many components. Some of these may be evaluated by Type A evaluation of measurement uncertainty from the statistical distribution of the quantity values from series of measurements and can be characterized by standard deviations. The other components, which may be evaluated by Type B evaluation of measurement uncertainty, can also be characterized by standard deviations, evaluated from probability density functions based on experience or other information.

uncertainty budget
statement of a measurement uncertainty, of the components of that measurement uncertainty, and of their calculation and combination.

Notes

An uncertainty budget should include the measurement model, estimates, and measurement uncertainties associated with the quantities in the measurement model, covariance, type of applied probability density functions, degrees of freedom, type of evaluation of measurement uncertainty, and any coverage factor.

input quantity in a measurement model
input quantity.

quantity that must be measured, or a quantity, the value of which can be otherwise obtained, in order to calculate a measured quantity value of a measurand.

influence quantity
quantity that, in a direct measurement, does not affect the quantity that is actually measured, but affects the relation between the indication and the measurement result.

Note

An indirect measurement involves a combination of direct measurements, each of which may be affected by influence quantities.

Example

Amount-of-substance concentration of bilirubin in a direct measurement of hemoglobin amount-of-substance concentration in human blood plasma.

output quantity in a measurement model
output quantity.

quantity, the measured value of which is calculated using the values of input quantities in a measurement model.

expanded measurement uncertainty
expanded uncertainty.

product of a combined standard measurement uncertainty and a factor larger than the number one.

Note

The factor depends upon the type of probability distribution of the output quantity in a measurement model and on the selected coverage probability.

coverage interval

interval containing the set of true quantity values of a measurand with a stated probability, based on the information available.

Note 1 A coverage interval should not be termed "confidence interval" to avoid confusion with the statistical concept.

Note 2 A coverage interval can be derived from an expanded measurement uncertainty.

coverage probability

probability that the set of true quantity values of a measurand is contained within a specified coverage interval.

Note

The coverage probability is also termed "level of confidence" in the GUM.

coverage factor

number larger than 1 by which a combined standard measurement uncertainty is multiplied to obtain an expanded measurement uncertainty.

Note

A coverage factor is symbolized k.

MISCELLANEA

sensitivity

sensitivity of a measuring system.

quotient of the change in an indication of a measuring system and the corresponding change in a value of a quantity being measured.

Note 1 Sensitivity of a measuring system can depend on the value of the quantity being measured.

Note 2 The change considered in a value of a quantity being measured must be large compared with the resolution.

resolution

smallest change in a quantity being measured that causes a perceptible change in the corresponding indication.

Note

Resolution can depend on, for example, noise (internal or external) or friction. It may also depend on the value of a quantity being measured.

detection limit

limit of detection.

measured quantity value, obtained by a given measurement procedure, for which the probability of falsely claiming the absence of a component in a material is β, given a probability α of falsely claiming its presence.

Note 1 IUPAC recommends default values for α and β equal to 0.05.

Note 2 The abbreviation LOD is sometimes used.

Note 3 The term "sensitivity" is discouraged for this concept.

verification

provision of objective evidence that a given item fulfils specified requirements.

Example 1

Confirmation that performance properties or legal requirements of a measuring system are achieved.

Example 2

Confirmation that a target measurement uncertainty can be met.

Note 1 When applicable, measurement uncertainty should be taken into consideration.

Note 2 The item may be, e.g., a process, measurement procedure, material, compound, or measuring system.

Note 3 Verification should not be confused with calibration. Not any verification is a validation.

validation

verification, where the specified requirements are adequate for an intended use.

Example

A measurement procedure, ordinarily used for the measurement of mass concentration of nitrogen in water, may be validated also for measurement in human serum.

interval

indication interval.

set of quantity values bounded by extreme possible indications.

Note

An indication interval is usually stated in terms of its smallest and greatest quantity values, for example "15-25 mL."

range of a nominal indication interval

absolute value of the difference between the extreme quantity values of a nominal indication interval.

Example

For a nominal indication interval of 15-25 mL, the range of the nominal indication interval is 10 mL.

Further Reading

There is an abundance of statistical text related to measurements in laboratories. Some textbooks that have appealed particularly to the author are listed below.

Several Internet sites provide useful information but are not always reliable.

Altman, Douglas G: Practical Statistics for Medical Research. Chapman and Hall, London 1991–1995, ISBN 0-412-27630-5.

Altman Douglas G, Machin David, Bryant Trevor N and Gardner Martin J. Statistics with confidence. British Medical Journal Publications 2nd Ed, Oxford. 2000. ISBN 978-0-7279-1375-3.

Armitage P, Berry G Matthews JNS: Statistical Methods in Medical Research. 4th Ed 2002. Blackwell Science Ltd Malden Mass, Oxford. ISBN 978-0-6320-5257-8.

Dybkaer R. An Ontology of Property for Physical, Chemical and Biological Systems. http://ontology.iupac.org. Accessed 2013-08-25

Engineering Statistics Handbook (NIST 2003). Several updates have been made since. http://www.itl.nist.gov/div898/handbook/. Accessed 2013-08-26.

Eurachem/CITAC Guide CG4 (QUAM 2000.1) Quantifying Uncertainty in Analytical Measurement. http://www.eurachem.org/guides/quam.htm.

Field, Andy Discovering Statistics using IBM SPSS Statistics. 4th edition SAGE Publications, Oxford. 2013 ISBN 978-1-4462-4917-8.

Heller Michel. Statement at The Templeton Prize News Conference, March 12, 2008. http://www.templetonprize.org/pdfs/heller_statement.pdf. Accessed 2013-06-30.

ISO 3534-1: 2006 Statistics—Vocabulary and symbols—Part 1: General Statistical Terms and Terms Used in Probability. Ed 2. Part 2: Applied statistics. Ed 2.

ISO 5725-2: Accuracy (Trueness and Precision) of measurement methods and results. Part 2. Basic method for the determination of repeatability and reproducibility of a standard measurement method. Geneva 1994, corr 2002.

IUPAC Compendium of Chemical Terminology—The Gold Book. http://goldbook.iupac.org/. Accessed 2013-08-26

Glanz Stanton A. Primer of Biostatistics. 4th Ed. McGraw Hill, New York, NY, USA 1996. ISBN 0-07-024268-2.

Kirkwood Betty R and Sterne Jonathan AC: Essential Medical Statistics. 2nd Ed. Blackwell Publishing Co. Malden, MA, USA 2003. ISBN 0-86542-871-9.

Miller JN and Miller JC: Statistics and Chemometrics for Analytical Chemistry, 6th Ed. Pearson Education, Harlow, UK. 2010. ISBN 978-0-273-73042-2.

Taylor John R: An Introduction to Error Analysis. The Study of Uncertainties in Physical Measurements. 2nd Ed. University Science Books, Sausalito, CA, USA, 1996. ISBN 978-0-935702-75-0.

Van Belle Gerald. Statistical Rules of Thumb. J. Wiley & Sons 2nd Ed Hoboken NJ, USA 2008. ISBN 978-0-470-14448-0.

Index

Note: Page numbers followed by *t* indicate tables.

Printed in the United States
By Bookmasters